THE NOISY OSCILLATOR

Random Mass, Frequency, Damping

Second Edition

THE NOISY OSCILLATOR

OSCILLATOR

Random Mass, Frequency, Damping

Second Edition

Moshe Gitterman
Bar-Ilan University, Israel

 **World Scientific**

NEW JERSEY · LONDON · SINGAPORE · BEIJING · SHANGHAI · HONG KONG · TAIPEI · CHENNAI

Published by

World Scientific Publishing Co. Pte. Ltd.

5 Toh Tuck Link, Singapore 596224

USA office: 27 Warren Street, Suite 401-402, Hackensack, NJ 07601

UK office: 57 Shelton Street, Covent Garden, London WC2H 9HE

British Library Cataloguing-in-Publication Data
A catalogue record for this book is available from the British Library.

THE NOISY OSCILLATOR
Random Mass, Frequency, Damping
(2nd Edition)

ISBN 978-981-4440-48-6

Printed in Singapore.

Preface to the Second Edition

The great interest in the noise oscillator stems from the fact that many different physical systems can be modeled by a harmonic oscillator. Indeed, any mass undergoing small oscillations around its equilibrium position, subject to a conservative force, behaves as a simple harmonic oscillator. Applications of this model include such diverse fields in physics as electrical circuits, liquid crystals, hydrodynamic systems, lasers, plasma, and nuclear reactions. Moreover, this model has also been applied to different phenomena in chemistry, biology, economics, and even the social sciences.

All phenomena in nature are subject to noise, that is, random disturbances that are usually considered undesirable. However, modern technology has learned to use such random disturbances, in the form of "stochastic resonance" and "deterministic chaos", for the amplification of weak signals.

This book deals with the effect of noise on various types of harmonic oscillators. Additive noise arises from non-zero temperatures or from very rapid dynamics of certain degrees of freedom. Multiplicative noise arises from the stochastic nature of external fields and boundary conditions. Noise may influence the coordinate or the velocity of the oscillator, leading to a "random frequency" or "random damping", respectively. The properties of a harmonic oscillator having a random frequency and/or random damping formed the content of the first edition of this book.

In the seven years since the first edition, a new type of stochastic oscillator has been investigated, namely, an oscillator having a random mass. The simplest example of this phenomenon involves the dynamics of a Brownian particle, subject to collisions with molecules from the surrounding medium which causes its well-known zig-zag motion. A stochastic element enters if the colliding molecules can also adhere to the Brownian particle for an arbitrary length of time. Such stochastic adhesion of the molecules leads to a random change in the mass of the Brownian particle.

The most important new material in the second edition is a detailed discussion of the effect on the motion of a harmonic oscillator due to a random mass, which is a new source of randomness. As will be shown in the text, the model of a harmonic oscillator with a random mass is not limited to Brownian particles, but has very wide application as an appropriate model in many areas of science. The analysis of an oscillator with a random mass is presented in two new chapters, entitled "Harmonic oscillator with random mass" and "Linear versus quadratic noise", The appearance of the second edition gave me the opportunity to correct some typographical errors, to include additional problems, and to alter the text in several places in the interest of greater clarity. This field continues to be the subject of intensive studies and eight new sections and an expanded bibliography were added to discuss the new results.

Preface to the First Edition

At one time I worked on the first heart sound produced by the left ventricular. The existence of a high frequency component was one of the predictions of the new theory, and I was very interested to get an experimental verification of this prediction. It was a reason for my talk for the group of cardiologists in one of the hospitals. I started my talk by saying: "Let us consider the left ventricular as a sphere or cylinder ... ". At the same moment I lost my audience who knew (as I did) that left ventricular has neither form of a sphere nor that of a cylinder. On the other hand, I knew how to build the model for the simplest geometry of sphere or cylinder. When an average physicists comes up against a new problem, he immediately starts to apply to this problem all the methods and ideas he knew from his previous experience[1].

The first step is the choice of some simplified model (geometric form of the left ventricular in my case). The simplest, the most general and the most used model is that of a harmonic oscillator. This model is used for the description of different phenomena in mechanics, optics, acoustics, electronics, engineering, etc. [1]. In fact, it has been applied everywhere, from quarks to cosmology. Moreover, a person who is worried by oscillations of prices in the stockmarket can be relaxed by classical music produced by the oscillations of string instruments.

Wandering about the Internet one can find many curious facts. Although the ancient Greeks already had a general idea of oscillations and used them in musical instruments, the first practical application of an oscillating system took place in 1602 by a physician in Venice named Santorio

---

[1]Of course, the great physicists do not come under these headings - recall A. Einstein with general relativity, P. Dirac with relativistic quantum mechanics, or L. Onsager with the two-dimensional Ising problem. In fact, they created not only the new physics, but also the new mathematics.

who had heard from his great friend Galileo about the general laws of the oscillations of a pendulum. Mr. Santorio called this system "pulsilogium" and used it to measure the pulse of his patients. Many other applications have been found in the last 400 years...

The two main features characterize this book. Firstly, the book contains a comprehensive description of all "oscillator-like" stochastic differential equations which were studied until 2012, and it can serve as a starting point for researches and engineers who meet these equation in their work. The second characteristic feature of this book is its simplicity and small volume.

Dozens of existing books and many hundreds of articles create a serious problem that confronts the author of a new book on stochastic differential equation. The problem consists of a difficult decision about restricting the consideration to some specific problem. My decision was to concentrate on a single stochastic one-dimensional classical oscillator, omitting thereby the quantum problem, interactions of oscillators and higher dimensions as well as the fascinating problem of deterministic chaos.

I decided not only to avoid all rigorous mathematical proofs and statements, but also to drop the large body of applications and the traditional comprehensive introduction to the mathematical theory of random processes. All necessary explanations are given in the appropriate sections of the book. If the material presented demands complicated calculations, I covered only the qualitative results referring for details to the original articles. Only the general knowledge of mathematical physics is required of the reader.

This book is devoted to "noisy" equations, i.e., equations which contain random forces, although for the reader's convenience, Chapter 2 contains a short review of the deterministic equations and the types of the second order (underdamped) differential equations involving additive and/or multiplicative noise which are considered, along with their simplified versions (first order differential equation) in great detail in this book. Chapter 3, which is devoted to the short general descriptions of noise, is required for understanding the ensuing material. In Chapter 4 we consider the Brownian motion, thereby paying respect to Einstein [2], Smoluchowski [3] and Langevin [4] who where the first to introduce a random force in deterministic equations for velocity and for coordinate. The following chapters contain detailed analyses of the overdamped and underdamped noisy oscillators.

Contents

Preface to the Second Edition   v

Preface to the First Edition   vii

1.   Deterministic and Random Oscillators   1

    1.1   Simple harmonic oscillator 1

    1.2   Damped harmonic oscillator 1

    1.3   Driven harmonic oscillator 2

    1.4   Driven, damped harmonic oscillator 2

    1.5   Non-linear oscillator in a double-well potential 3

    1.6   Non-linear oscillator in a single-well potential 3

    1.7   Harmonic oscillator with external noise 3

    1.8   Brownian motion 4

    1.9   Harmonic oscillator with random frequency 4

    1.10   Harmonic oscillator with random damping 5

    1.11   Harmonic oscillator with random mass 5

2.   White and Color Noise   9

    2.1   Dichotomous noise 10

    2.2   Trichotomous noise 11

    2.3   Polynomials dichotomous noise 11

    2.4   White Poisson noise 12

    2.5   Shot noise 13

3.   Brownian Motion   15

    3.1   In the beginning... 15

    3.2   Fokker-Planck equation 18

		3.2.1	Additive white noise	18
		3.2.2	Multiplicative white noise	19
	3.3		Active Brownian particle	21
	3.4		Color noise: unified color noise approximation (UCNA)	22
	3.5		Brownian motion and anomalous diffusion	24
	3.6		Brownian motion in a fluctuating medium	25
	3.7		Brownian motion near the critical point	26

4. Overdamped Harmonic Oscillator with Additive Noise   29

| | 4.1 | | Additive white noise | 30 |
| | 4.2 | | Additive noise and periodic force | 30 |

5. Overdamped Harmonic Oscillator with Multiplicative Noise   31

	5.1		Multiplicative noise (shift of stable points)	31
	5.2		Multiplicative and additive noises	32
		5.2.1	Two white noises	32
		5.2.2	Two correlated white noises	33
		5.2.3	Two correlated dichotomous noises	34
	5.3		Multiplicative color noise and periodic signal (stochastic resonance (SR) in linear systems)	35
	5.4		Stochastic resonance in a overdamped system with signal-modulated noise	38

6. Overdamped Single-well Oscillator   41

	6.1		Steady state	41
		6.1.1	White noises	41
		6.1.2	Multiplicative noise (gene selection)	43
		6.1.3	Dichotomous noise	44
		6.1.4	Poisson white noise	45
	6.2		Response to a periodic force (noise-enhanced stability)	45
	6.3		Piece-wise model of a metastable state	49
	6.4		Rectangular potential barrier (stabilization of metastable state)	51

7. Overdamped Double-well Oscillator   59

	7.1		Steady state	59
		7.1.1	White noise	59
		7.1.2	Dichotomous noise	61

7.2   Eigenfunction expansion of the Fokker-Planck equation   .   62

7.3   Matrix continued fraction method   63

7.4   Mean first-passage time   65

7.5   Response to a periodic force (stochastic resonance)   67

7.6   Mechanism of stochastic resonance   71

7.7   Fluctuating potential barrier (resonance activation)   73

     7.7.1   Piece-wise linear potential   73

     7.7.2   Phenomenological model   75

     7.7.3   Coherent stochastic resonance   76

8.   Harmonic Oscillator with Additive Noise   77

    8.1   Internal and external noise   77

    8.2   White and dichotomous noise   77

    8.3   Additive noise and parametric oscillations   79

9.   Nonlinear Oscillator with Additive Noise   81

    9.1   Statistical linearization   81

    9.2   Double-well oscillator with additive noise   82

    9.3   Double-well oscillator driven by two periodic fields

        (vibrational resonance)   83

    9.4   Stochastic resonance in linear system subject to two periodic fields and random forces   85

10.   Harmonic Oscillator with Random Frequency   87

    10.1   First moment for the random frequency   87

       10.1.1   Force-free oscillator   87

       10.1.2   White noise   87

       10.1.3   Color noise   87

    10.2   Second moment for a random frequency   88

    10.3   Maxwell equation with random dielectric constant   90

       10.3.1   Driven Maxwell equation   92

    10.4   Stochastic resonance in the oscillator with random frequency   93

11.   Harmonic Oscillator with Random Damping   97

    11.1   First moment for random damping   97

       11.1.1   Force-free oscillator   97

    11.2   Second moment for random damping   99

11.3   Force-free oscillator 102
        11.3.1   White noise 102
        11.3.2   Dichotomous noise 102
        11.3.3   Poisson noise 102
        11.3.4   Driven oscillator 103
11.4   Second moments 103
11.5   Correlation functions 104
11.6   Stochastic resonance in the oscillator with random
        damping 105
11.7   Periodically varying damping 108

12. Linear vs Quadratic Noise   111

12.1   Brownian motion 112
12.2   Harmonic oscillator with random frequency 113
12.3   Harmonic oscillator with random damping 115
12.4   Harmonic oscillator with random mass 117
        12.4.1   Linear noise of small strength $(\sigma^2 < 1)$ 117
        12.4.2   Linear-response theory 119
        12.4.3   Quadratic noise 120

13. Nonlinear Oscillator with Multiplicative Noise   123

13.1   Double-well potential (noise induced reentrant
        transition) 123
13.2   Duffing oscillator 126
13.3   Van der Pol oscillator 127

14. Harmonic Oscillator with Random Mass   131

14.1   Basic equations 133
14.2   First moment 136
14.3   White noise 137
14.4   Symmetric dichotomous noise 138
14.5   Asymmetric dichotomous noise 140
14.6   Second moment 140
14.7   Stochastic resonance in the oscillator with random
        mass 143
14.8   Stability conditions 147
14.9   White noise 150
14.10   Dichotomous noise 152

14.11 Resonance phenomena 154

      14.11.1 Vibrational resonance 156

      14.11.2 Deterministic chaos ("Erratic" behavior) 159

15. In the Future...   163

Bibliography   165

Index   173

Chapter 1

Deterministic and Random Oscillators

This chapter reviews all types of differential equations which will be analyzed later on in more detail. We present here the second order (underdamped) differential equations. The first-order (overdamped) equations, which we will consider first can be obtained from it by leaving only the first derivative in these equations.

1.1 Simple harmonic oscillator

Such an oscillator is a simple system where the net force is directly proportional to the displacement x of the mass m from the equilibrium position $x = 0$ and pointing in the opposite direction (the Hook's law). According to Newton's law of motion,

$$m\frac{d^2x}{dt^2} = -kx \quad \text{or} \quad m\frac{d^2x}{dt^2} + kx = 0. \tag{1.1}$$

The solution of this equation has a form

$$x = C\cos\left(\omega_0 t + \phi\right), \tag{1.2}$$

where the angular frequency $\omega_0 = \sqrt{k/m}$ depends only upon the parameters of the system, while the amplitude A and the phase ϕ are constants determined by the initial displacement $x\left(t = 0\right)$ and velocity (dx/dt) at $t = 0$.

1.2 Damped harmonic oscillator

The preceding analysis can be supplemented by a dissipative force which points in the opposite direction and is usually assumed to be proportional to the velocity,

$$m\frac{d^2x}{dt^2} = -kx - 2\gamma\frac{dx}{dt} \quad \text{or} \quad m\frac{d^2x}{dt^2} + 2\gamma\frac{dx}{dt} + kx = 0. \tag{1.3}$$

The solution of Equation (1.3) for $k > \gamma$ has the form

$$x = C\exp\left(-\frac{\gamma t}{m}\right)\cos\left(\omega_1 t + \phi\right), \tag{1.4}$$

with $\omega_1 = \sqrt{\omega_0^2 - (\gamma/m)^2}$.

1.3 Driven harmonic oscillator

Another generalization of the equations for a simple harmonic oscillator (1.1) and a damped harmonic oscillator (1.3) is one driven by some time-dependent external force, the simplest form of which is a periodic function of time, i.e.,

$$m\frac{d^2x}{dt^2} + kx = A\cos\left(\Omega t + \beta\right). \tag{1.5}$$

This has a driven solution of the form

$$x = \frac{A}{m\left(\omega_0^2 - \Omega^2\right)}\cos\left(\Omega t + \beta\right). \tag{1.6}$$

When the external frequency Ω approaches the intrinsic frequency ω_0, the steady state amplitude approaches infinity (dynamic resonance). A classic demonstrations of this dynamic resonance are two architectural flaws in the US. The first is the Takoma bridge which was destroyed by the wind's force with the resonance frequency, and the second one is the Paramount Communication building in New York transformed into luxury apartments building where due to the winds the top floors were twisted and windows were prying loose from their casements.

1.4 Driven, damped harmonic oscillator

On combining the damping and driving forces, one arrives at the following equation

$$m\frac{d^2x}{dt^2} + kx + 2\gamma\frac{dx}{dt} = A\cos\left(\Omega t + \beta\right). \tag{1.7}$$

The general solution of Equation (1.7) is a sum of a transient and a steady state solutions

$$x = C \exp\left(-\frac{\gamma t}{m}\right) \cos\left(\omega_1 t + \phi\right) +$$

$$\frac{A}{m\sqrt{\left(\omega_0^2 - \Omega^2\right)^2 + \frac{4\gamma^2\Omega^2}{m^2}}} \sin\left(\Omega t + \phi + \beta\right), \tag{1.8}$$

with $\beta = \tan^{-1}\left[\left(\omega_0^2 - \Omega^2\right)/2\gamma\Omega\right]$.

1.5 Non-linear oscillator in a double-well potential

Up to now, we have considered the linear oscillator with the restoring force f proportional to the displacement x, $f \approx -kx$. In more general case this dependence can be nonlinear, $f \approx ax - bx^n$. Hereafter, we restrict our attention to the most interesting cases $n = 2$ and $n = 3$.

The linear force in Equations (1.1), (1.3), (1.7) corresponds to the potential $U = kx^2/2$. For the more complicated case of a double-well potential energy $U = -ax^2/2 + bx^4/4$, with two minima located at $x = \pm\sqrt{a/b}$ and a maximum at $x = 0$, the equation of motion has the form

$$m\frac{d^2x}{dt^2} - ax + bx^3 = 0. \tag{1.9}$$

1.6 Non-linear oscillator in a single-well potential

Another much used form of the potential energy is one of the form $U = -ax^2/2 + bx^3/3$, with an unstable minimum at $x = a/b$, which results in the following equation of motion

$$m\frac{d^2x}{dt^2} - ax + bx^2 = 0. \tag{1.10}$$

Both Equations (1.9) and (1.10) can be easily generalized to include a damping term and an external force. However, one cannot find the exact solutions of these equations even for a simplified form of these equations .

1.7 Harmonic oscillator with external noise

In the foregoing we considered a pure mechanical system (zero temperature). However, for all finite temperatures the dynamic Equations (1.1),

(1.4) have to be supplemented by thermal noise, i.e.,

$$m\frac{d^2x}{dt^2} + 2\gamma\frac{dx}{dt} + kx = \xi(t), \qquad (1.11)$$

where $\xi(t)$ is the random variable with zero mean $\langle\xi(t)\rangle = 0$ and the variance $\langle\xi^2(t)\rangle$ which for the thermal noise must satisfy the fluctuation-dissipation theorem [5], $\langle\xi^2(t)\rangle = 2\gamma\kappa T$, where κ is the Boltzmann constant. The noise acting on a system can be also external (and not thermal noise) with no special requirement for the value of $\langle\xi^2(t)\rangle$. Still another way to justify the validity of Equation (1.11) is as follows: in considering only one (slow) mode $x(t)$ of a complex system, one may take into account the influence of other (fast) modes by introducing a random force into the dynamic equations.

1.8 Brownian motion

Equation (1.11) with $m = 0$ and x meaning the particle's velocity v describes Brownian motion, where the force acting on the Brownian particle consists of the systematic force $-kv$ and the random force $\xi(t)$,

$$\gamma\frac{dv}{dt} + kv = \xi(t). \qquad (1.12)$$

Note, that the caution is required in going from the oscillator equation to that of the Brownian particle, since the simple substitution of $m \to 0$ in Equation (1.11) will decrease the order of the differential equation, and one has to use the singular perturbation theory [6].

1.9 Harmonic oscillator with random frequency

The random force $\xi(t)$ enters Equation (1.11) additively. When the noise has an external origin rather than an internal one, for instance if it arises from the fluctuations of the potential energy $U = kx^2[1 + \xi(t)]/2$, the equation of motion of an oscillator with an external periodic force $A\sin(\Omega t)$, will take the following form:

$$m\frac{d^2x}{dt^2} + 2\gamma\frac{dx}{dt} + k[1 + \xi(t)]x = A\sin(\Omega t). \qquad (1.13)$$

The internal dynamics of a force-free harmonic oscillator ($A = 0$) with random frequency is a subject that has been extensively investigated in different fields, including physics (on-off intermittence [7], dye lasers [8], wave propagation in random media [9], turbulent flows [10]), biology (population dynamics [11]), economics (stock market prices [12]) and so on.

Note that for $A = \gamma = 0$ and for x and t being replaced by the electric field and coordinate, respectively, Equation (1.13) transforms into the Maxwell equation with a random dielectric constant which has also been intensively studied [13]. We considered Equation (1.13) with an external force [14].

1.10 Harmonic oscillator with random damping

Another possibility for the generalization of the dynamic Equation (1.7) is the incorporation of random damping

$$m\frac{d^2x}{dt^2} + 2\gamma\left[1 + \xi\left(t\right)\right]\frac{dx}{dt} + kx = A\sin\left(\Omega t\right). \tag{1.14}$$

The first time this equation (with $A = 0$) was used [15] for the problem of water waves influenced by a turbulent wind field. However, this equation (with the coordinate x and time t replaced by the order parameter and coordinate, respectively) is transformed into the Ginzburg-Landau equation with a convective term which describes phase transitions in moving systems [16]. There are an increasing number of problems where the particles advected by the mean flow pass through the region under study. These include problems of phase transition under shear [17], open flows of liquids [18], Rayleigh-Benard and Taylor-Couette problems in fluid dynamics [19], dendritic growth [20], chemical waves [21], and the motion of vortices [22].

Notice the important difference between the linear Equation (1.11) with an additive noise and Equations (1.13), (1.14) with multiplicative noise. The latter show, in fact, "quasi-nonlinear behavior" including the stochastic resonance phenomenon [14], [23].

1.11 Harmonic oscillator with random mass

We recently studied [24] still another possibility for introducing randomness in the oscillator Equation (1.3), by considering an oscillator with a random mass, which describes a new type of Brownian motion - Brownian motion with adhesion. In this situation the molecules of the surrounding medium not only randomly collide with the Brownian particle, which produces its well-known zigzag motion, but they also stick to the Brownian particle for some (random) time, thereby changing the mass of the Brownian particle. The appropriate equation of motion has the following form

$$m\left[1 + \xi\left(t\right)\right]\frac{d^2x}{dt^2} + \nu\frac{dx}{dt} + kx = m\eta\left(t\right). \tag{1.15}$$

One additional comment is necessary. The change of the mass leads to the change of the momentum. However, in the closed system (Brownian particle and surrounded molecules) the momentum is conserved. We do not consider the detail behavior of the surrounding molecules, which are in charge for the conservation law, restricting our analysis to the random collisions of these molecules with the Brownian particle.

There are many situations in which chemical and biological solutions contain small particles which are capable not only of colliding with a large particle, but they may also adhere to it. The diffusion of clusters with randomly growing masses was also considered [25]. There are also some applications of a variable-mass oscillator [26]. Modern applications of such a model include a nano-mechanical resonator which randomly absorbs and desorbs molecules [27]. Among other applications of (1.15) is an RLC electrical circuit subject to a voltage $V\left(t\right)$ with a fluctuating inductance L, which is described by the following equation

$$\left[L + \xi\left(t\right)\right]\frac{d^2J}{dt^2} + R\frac{dJ}{dt} + \frac{1}{C}J = \frac{dV}{dt}. \tag{1.16}$$

Equations (1.15) and (1.16) describe the dynamic equation with a mass that can both increase or decrease due to fluctuations. As distinct from Equation (1.15), we here consider in Equation (1.17) a positive random force $\xi^2\left(t\right)$, which corresponds to the fact that the mass of the Brownian particle can only increase due to the adhesion of the molecules of the surrounding medium,

$$m\left[1 + \xi^2\left(t\right)\right]\frac{d^2x}{dt^2} + \nu\frac{dx}{dt} + kx = m\eta\left(t\right) \tag{1.17}$$

which can be rewritten as

$$\left[1 + \xi^2\left(t\right)\right]\frac{d^2x}{dt^2} + \gamma\frac{dx}{dt} + \omega^2x = \eta\left(t\right) \tag{1.18}$$

where $\gamma = \nu/m$ is the damping coefficient and $\omega = \sqrt{k/m}$ is the eigenfrequency.

In order to reduce the usual equation of motion of the Brownian particle to that of a random harmonic oscillator, we consider Brownian motion in a parabolic potential which leads to a linear x term in Equations (1.15) and (1.18).

The forced harmonic oscillator is described by Equation (1.18) with the random force $\eta(t)$ replaced by an additional periodic force $a \sin(\Omega t)$,

$$\left([1 + \xi^2(t)]\frac{d^2x}{dt^2} + \gamma\frac{dx}{dt} + \omega^2 x = a \sin(\Omega t)\right). \tag{1.19}$$

The simplest form of the color noise $\xi(t)$ in (1.18) and (1.19) is the asymmetric dichotomous noise (random telegraphic process), which means that the random variable $\xi(t)$ takes values A or $-B$. Denote the rate of transition for A to $-B$ by τ_1, and the reverse rate by τ_2. For this form of the Ornstein-Uhlenbeck noise, the correlation function is

$$\langle\xi(t_1)\xi(t_2)\rangle = \sigma^2/\lambda \exp -\lambda|t_1 - t_2|; \qquad \lambda = \frac{1}{\tau_1 + \tau_2}. \tag{1.20}$$

We choose the parameters to make $\langle\xi(t)\rangle = 0$, which implies the relation

$$\tau_1 A = \tau_2 B. \tag{1.21}$$

The quadratic noise $\xi^2(t)$ can be written as

$$\xi^2(t) = \sigma^2 + \Delta\xi \tag{1.22}$$

with $\sigma^2 = AB$ and $\Delta = A - B$. Indeed, for $\xi = A$, one obtains $\xi^2 = AB + (A - B)A = A^2$, and for $\xi = B$, $\xi^2 = B^2$. Inserting (1.22) into (1.19) gives

$$(1 + \sigma^2 + \Delta\xi)\frac{d^2x}{dt^2} + \gamma\frac{dx}{dt} + \omega^2 x = a \sin(\Omega t) \tag{1.23}$$

Multiplying Eq (1.23) with $a = 0$ by $1 + \sigma^2 - \Delta\xi(t)$, one obtains,

$$(T - \Delta^3\xi)\frac{d^2x}{dt^2} + (1 + \sigma^2)\left(\gamma\frac{dx}{dt} + \omega^2 x\right) = \Delta\xi(t)\left(\gamma\frac{dx}{dt} + \omega^2 x\right) \tag{1.24}$$

where $T = (1 + \sigma^2)^2 - \Delta^2\sigma^2$.

Equation (1.18) with $\eta = 0$ and $1 + \xi^2(t)$ replaced by $1 + \sigma^2 + \Delta\xi$ can be rewritten as

$$L(x) = -\Delta\xi\frac{d^2x}{dt^2}; \qquad L(x) = (1 + \sigma^2)\frac{d^2x}{dt^2} + \gamma\frac{dx}{dt} + \omega^2 x. \tag{1.25}$$

Chapter 2

White and Color Noise

In the following we will consider noise $\xi(t)$ with $\langle \xi(t) \rangle = 0$ and the correlator

$$\langle \xi(t_1) \xi(t_2) \rangle = r |t_1 - t_2| \equiv r(z). \tag{2.1}$$

Two integrals of (2.1) characterize the fluctuations : the strength of the noise D

$$D = \int_0^\infty \langle \xi(t) \xi(t+z) \rangle \, dz, \tag{2.2}$$

and the correlation time τ,

$$\tau = \frac{1}{D} \int_0^\infty z \langle \xi(t) \xi(t+z) \rangle \, dz. \tag{2.3}$$

Traditionally one considers two different forms of noise, white and colored noise. For white noise the function $r |t_1 - t_2|$ has the form of a delta-function,

$$\langle \xi(t_1) \xi(t_2) \rangle = D\delta(t - t_1), \tag{2.4}$$

The name "white" noise comes from the fact that the Fourier transform of (2.4) is "white" being constant without any characteristic frequency. Equation (2.4) means that noises $\xi(t_1)$ and $\xi(t_2)$ are statistically independent, no matter how close are t_1 and t_2. This extreme assumption, which leads to non-physical infinite value of $\langle \xi^2(t) \rangle$ in (2.4), means, in fact, that the correlation time τ is not zero as it assumed in (2.4), but smaller than all the other characteristic times in a problem. We will return to this subject in the next chapter on examining the Fokker-Planck equation for multiplicative noise.

All other non-white noises are called colored noise. A widely used noise is the Ornstein-Uhlenbeck exponentially correlated noise which can be written in two forms,

$$\langle \xi(t)\,\xi(t_1)\rangle = \frac{\sigma^2}{\lambda} \exp\left(-\lambda\,|t-t_1|\right), \qquad (2.5)$$

or

$$\langle \xi(t)\,\xi(t_1)\rangle = D\exp\left(-\frac{|t-t_1|}{\tau}\right). \qquad (2.6)$$

White noise (2.4) is defined by its strength D while the Ornstein - Uhlenbeck noise is characterized by two parameters, λ and σ^2, or τ and D. The transition from the Ornstein -Uhlenbeck noise to white noise (2.4) is performed by the limit $\tau \to 0$ in (2.6), or by $\sigma^2 \to \infty$ and $\lambda \to \infty$ in (2.5) in such a way that $\sigma^2/\lambda = D$.

A slightly generalized form of the Ornstein-Uhlenbeck noise is the so-called narrow-band colored noise with a correlator of the form,

$$\langle \xi(t)\,\xi(t_1)\rangle = \sigma^2 \exp -\lambda\,|t-t_1| \ \cos\left(\omega_2\,|t-t_1|\right). \qquad (2.7)$$

There are different forms of colored noises which we will briefly review in this chapter.

2.1   Dichotomous noise

A special type of colored noise with which we shall be concerned is the symmetric dichotomous noise (random telegraph signal) where the random variable $\xi(t)$ may take one of the values $\xi = \pm\sigma$ with the mean waiting time $(\lambda/2)^{-1}$ in each of these two states. Like the Ornstein–Uhlenbeck noise, dichotomous noise is characterized by the correlators (2.5)-(2.6).

In what follows we will use the Furutzu-Novikov-Shapiro-Loginov procedure [28], [29] for splitting the higher order correlations, which for exponentially correlated noise yields

$$\frac{d}{dt}\langle \xi \cdot g\rangle = \left\langle \xi\frac{dg}{dt}\right\rangle - \lambda\langle \xi \cdot g\rangle, \qquad (2.8)$$

where g is some function of noise, $g = g\{\xi\}$. If $\frac{dg}{dt} = B\xi$, then Equation (2.8) becomes

$$\frac{d}{dt}\langle \xi \cdot g\rangle = B\langle \xi^2\rangle - \lambda\langle \xi \cdot g\rangle, \qquad (2.9)$$

and for white noise ($\xi^2 \to \infty$ and $\lambda \to \infty$ with $\xi^2/\lambda = 2D$), one gets

$$\langle \xi \cdot g\rangle = 2BD. \qquad (2.10)$$

2.2 Trichotomous noise

Trichotomous noise offers the simple generalization of the dichotomous noise and consists of random jumps between three values, $\pm a$ and zero. This model was introduced by R. Mankin and collaborators [30] and succesfully applied to an oscillator with random frequency. In Chapter 14 we will apply this model to an oscillator with random mass.

For the stationary states, the probabilities P of values $\pm a$ and zero are

$$P(-a) = P(a) = q; \qquad P(0) = 1 - 2q. \tag{2.11}$$

The supplementary conditions to the Ornstein-Uhlenbeck correlations (2.5) gives

$$\xi^3 = a^2\xi; \qquad \langle \xi^2 \rangle = 2qa^2. \tag{2.12}$$

Another peculiar feature of trichotomous noise is the splitting of the correlation, which we usully perform by the use of the Shapiro-Loginov procedure, which, however, has to be modiified compared with Equation (2.8). Indeed, in Equation (2.8) we use $\langle \xi \rangle = 0$ while for the trichotomous noise inserting $\widetilde{\xi} = \xi^2 - 2qa^2$ into (2.8) for $g = x$ and $g = y$ results in

$$\frac{d}{dt} \langle \xi^2 x \rangle = \langle \xi^2 y \rangle - \lambda \langle \xi^2 x \rangle + 2q\lambda a^2 \langle x \rangle \tag{2.13}$$

$$\frac{d}{dt} \langle \xi^2 y \rangle = \left\langle \xi^2 \frac{dy}{dt} \right\rangle - \lambda \langle \xi^2 y \rangle + 2q\lambda a^2 \langle y \rangle =$$
$$- a^2 \left(\frac{d}{dt} + \lambda \right) \langle \xi y \rangle - (\gamma + \lambda) \langle \xi^2 y \rangle -$$
$$\omega^2 \langle \xi^2 x \rangle + 2q\lambda a^2 \langle y \rangle + 2qAa^2 \sin(\Omega t). \tag{2.14}$$

2.3 Polynomials dichotomous noise

The polynomial dichtomous noise is the natural generalization of dichotomous noise. We consider it by the example of an oscillator with random mass, which will be fully considered in Chapter 14. Here we use the most general case of polynomial dichotomous noise [31], which transforms the oscillator equation to the following form

$$\left[1 + \sum_{k=1}^{n} a_k \xi^k(t) \right] \frac{d^2x}{dt^2} + \gamma \frac{dx}{dt} + \omega^2 x = \eta(t) \tag{2.15}$$

where $\eta(t)$ is white noise

$$\langle \eta(t_1) \eta(t_2) \rangle = D_1 \delta(t_1 - t_2). \tag{2.16}$$

For asymmetric dichotomous noise $\xi(t)$, one gets [31]

$$\left\langle \xi^k(t) g(t) \right\rangle = b_k \langle \xi(t) g(t) \rangle + c_k \langle g(t) \rangle \tag{2.17}$$

where

$$b_k = \frac{A^k - (-B)^k}{A + B}; \qquad c_k = \frac{BA^k + A(-B)^k}{A + B}. \tag{2.18}$$

It is easy to check that for $k = 2$, Equation (2.17) reduces to Equation (1.22), after multiplying the latter equation by $g(t)$ and averaging.

2.4   White Poisson noise

We define the white Poisson noise as a train of δ-functions at random times t_i, i.e.

$$\xi(t) = \sum_{i=1}^{n(t)} \omega_i \delta(t - t_i) - \beta \langle \omega \rangle \tag{2.19}$$

where $n(t)$ is the Poisson process with parameter β which controls the average time difference between two pulses. The amplitude of the pulses ω_i is exponentially distributed with the mean value ω,

$$\rho(\omega_i) = \frac{\exp(-\omega_i/\omega_0)}{\omega_0}. \tag{2.20}$$

Here β is the mean density of t_i, and the last term in (2.19) is added to insure a zero value for the mean noise, $\langle \xi(t) \rangle = 0$. In order to find the connection between Poisson noise and asymmetric dichotomous noise, one has to find the connection between β and ω_0 and the parameters A and the mean time τ between two states A and $-B$ of asymmetric dichotomous noise. Since parameters β^{-1} and τ^{-1} define the mean time between two successive impulses and the mean time in the state A, they are equal, $\beta = \tau$. The correlation function of the noise (2.19) has the following form [32].

$$\langle \xi(t_1) \xi(t_2) \rangle = 2\beta \omega_0^2 \delta(t_2 - t_1). \tag{2.21}$$

Comparing (2.21) with the correlation function (2.4) of dichotomous noise, one concludes that

$$D \equiv \frac{\sigma^2}{\lambda} \to 2\beta \omega_0^2. \tag{2.22}$$

Contrary to Gaussian white noise, which fully defined by its strength D, Poisson noise is defined by D and additional non-Gaussian parameter ω_0. Transition from white to Poisson noise is affected as

$$\sigma^2 \to 2\beta\omega_0^2\lambda; \qquad \lambda \to \omega_0\lambda. \tag{2.23}$$

2.5 Shot noise

The generalization of (2.19) results in the so-called shot noise,

$$\xi(t) = \sum_i \omega_i f(t - t_i) - g(t) \tag{2.24}$$

where ω_i and t_i are random variables described by the generalized Poisson process (2.24), where $f(t)$ is some response function, and $g(t)$ is the so-called deterministic compensator responsible for the condition $\langle\xi(t)\rangle = 0$.

If the "width" $t - t_i$ of $f(t - t_i)$ is small compared with other characteristic times of a problem, one can simplify Equation (2.24) by replacing $f(t - t_i)$ by the $\delta-$ function (white shot noise),

$$\xi(t) = \sum_i \omega_i \delta(t - t_i) - g(t). \tag{2.25}$$

Chapter 3

Brownian Motion

3.1 In the beginning...

This name came from the observation of peculiar erratic motion of a meso-scopic particle of a micron size floating among microscopic particles performed by Robert Brown in 1827 [33]. The theoretical description of this phenomenon was given by Einstein [2], Smoluchowski [3], and Langevin [4] who introduced the molecular-kinetics corrections to deterministic equations. Langevin considered the equation of motion for the position of the Brownian particle while Einstein and Smoluchowski that for its velocity[1]. They assume that the total force acting on the Brownian particle is decomposed into a systematic force (viscous friction proportional to velocity, $f = -\gamma v$), and a fluctuation force $\xi(t)$ exerted on the Brownian particle by the molecules of the surrounding medium. This fluctuation force comes from the different number of molecular collisions with a Brownian particle from two opposite sites, resulting in its random motion. Then, for a one-dimensional system Newton's law gives

$$m\frac{dv}{dt} = -\gamma v + \xi(t). \tag{3.1}$$

The simplest assumption of white noise (2.4) has been made in the original articles.

Both terms in the right-hand side of Equation (3.1) are connected with collisions between a Brownian particle and molecules of the surrounding media. The separation of this interaction into two parts is somewhat artificial, and it is not surprising, therefore, that their characteristics are connected by the fluctuation-dissipation theorem [5].

---

[1] For the velocity they used the twenty four-letters German word "Bewegungs-geschwindinkeit"!

Equation (3.1) has a simple solution

$$v = v(t = 0) \exp\left(-\frac{\gamma t}{m}\right) + \frac{1}{m}\int_0^t \exp\left[-\frac{\gamma}{m}(t - t_1)\right] \xi(t_1)\, dt_1. \quad (3.2)$$

Integration of (3.2) over t gives

$$x = x(t = 0) + \frac{mv(t = 0)}{\gamma}\left[1 - \exp\left(-\frac{\gamma t}{m}\right)\right] +$$

$$\frac{1}{\gamma}\int_0^t \left\{1 - \exp\left[-\frac{\gamma}{m}(t - t_1)\right]\right\} \xi(t_1)\, dt_1. \quad (3.3)$$

Using (3.3) one obtains for the mean-square displacement

$$\langle \sigma^2 \rangle \equiv \left\langle [x(t) - x(t = 0)]^2 \right\rangle = \frac{m^2 v(t = 0)^2}{\gamma^2}\left[1 - \exp\left(-\frac{\gamma t}{m}\right)\right]^2 +$$

$$\frac{1}{\gamma^2}\int_0^t dt_1 \int_0^t dt_2 \left\{1 - \exp\left[-\frac{\gamma(t - t_1)}{m}\right]\right\} \times$$

$$\left\{1 - \exp\left[-\frac{\gamma(t - t_2)}{m}\right]\right\} \langle \xi(t_1)\xi(t_2)\rangle. \quad (3.4)$$

Substituting in (3.4) the correlator (2.4) of white noise, one gets

$$\langle \sigma^2 \rangle = \frac{2Dt}{\gamma^2} + \frac{m^2 v(t = 0)^2}{\gamma^2}\left[1 - \exp\left(-\frac{\gamma}{m}t\right)\right]^2 +$$

$$\frac{mD}{\gamma^3}\left[-3 + 4\exp\left(-\frac{\gamma}{m}t\right) - \exp\left(-\frac{2\gamma}{m}t\right)\right] \quad (3.5)$$

and for long times, $t \to \infty$,

$$\left\langle (\Delta x)^2 \right\rangle = \frac{2Dt}{\gamma^2} = \frac{2\varkappa T}{\gamma}t. \quad (3.6)$$

The last equality in (3.6) follows from the fluctuation-dissipation theorem [5], $\langle \xi^2(t)\rangle = 2\gamma\kappa T$, where κ is the Boltzmann constant. The latter simply means that the power entering the system from the external random force must be entirely dissipated and given off to the thermostat in order that the equilibrium state of the system should not be disturbed. Analogously, one gets from (3.2) for the long time limit value of the mean velocity

$$\langle v^2 \rangle = \frac{\varkappa T}{m}, \quad (3.7)$$

the well-known statistical mechanics equipartition theorem. The equilibrium distribution comes about as balance of two contrary processes in Equation (3.1), the random force which tends to increase the velocity of the Brownian particle, and the damped force which tries to stop the particle.

Till now we considered the white noise in the dynamic Equation (3.1). We consider now the more general color noise, which we will take in the simplest form of dichotomous noise (random telegraph signal), which randomly jumps between two values, $\pm\sigma$. Such noise is characterized by the Ornstein-Uhlenbeck correlation function (1.20).

The use of dichotomous noise offers a major advantage over other types of color noise by terminating an infinite set of higher-order correlators, using the fact that $\langle \xi^2(t) \rangle = \sigma^2$.

Multiplying Equation (3.1) with ξ replaced by η by $2v$ and averaging, one obtains for stationary states ($d/dt... = 0$),

$$\langle v^2 \rangle = \frac{1}{2\gamma} \langle \eta v \rangle. \tag{3.8}$$

Multiplying Equation (3.1) by η and averaging,

$$\left\langle \eta \frac{dv}{dt} \right\rangle + 2\gamma \langle \eta v \rangle = \langle \eta^2(t) \rangle. \tag{3.9}$$

For the splitting of the correlators in Equation (3.9), we use the well-known Shapiro-Loginov procedure (2.8), which yields for exponentially correlated noise. Combining Equations (3.8) and (3.9), one obtains for the stationary state,

$$\langle \eta v \rangle = \frac{\sigma^2}{(2\gamma + \lambda)}. \tag{3.10}$$

From Equations (3.8) and (3.10), one gets

$$\langle v^2 \rangle = \frac{\sigma^2}{2\gamma(2\gamma + \lambda)}. \tag{3.11}$$

Equation (3.11) has been obtained for color noise (1.20) for $t \to \infty$. For white noise, as we have seen in (3.6), $\langle v^2 \rangle \to \infty$ directly proportional to t.

In the following, we will use also other forms of splitting [29],

$$\frac{d}{dt} \langle \xi \eta \phi \rangle = \left\langle \xi \eta \frac{d\phi}{dt} \right\rangle - \lambda \langle \xi \eta \phi \rangle \tag{3.12}$$

and

$$\frac{d}{dt} \langle \xi^2 \phi \rangle = \left\langle \xi^2 \frac{d\phi}{dt} \right\rangle - 2\lambda \langle \xi^2 \phi \rangle + 2\sigma^2 \langle \phi \rangle, \tag{3.13}$$

where ϕ is an arbitrary function of ξ. For white noise $\eta(t)$ and $g = v$ or $g = x$ with $dv/dt = C - \Delta\xi dv/dt - 2\gamma v + \eta$ and $dx/dt = v$, Equation (2.8) gives

$$\left(\frac{d}{dt} + \lambda + 2\gamma \right) \langle \eta v \rangle = D - \Delta \left\langle \xi \eta \frac{dv}{dt} \right\rangle; \quad \left(\frac{d}{dt} + \lambda \right) \langle \eta x \rangle = \langle \eta v \rangle. \tag{3.14}$$

Using Equation (3.12) with $\phi = y$, one rewrites the last equations for stationary states ($d/dt... = 0$),

$$(\lambda + 2\gamma) \langle \eta v \rangle = D - \Delta\lambda \langle \xi \eta v \rangle; \quad \lambda \langle \eta x \rangle = \langle \eta v \rangle. \tag{3.15}$$

3.2 Fokker-Planck equation

3.2.1 *Additive white noise*

The stochastic Equation (3.1) which is called Langevin equation, describes the motion of an individual Brownian particle. The random force $\xi(t)$ in this equation causes the solution of this equation $v(t)$ to be random as well. Equivalently, one can consider an ensemble of Brownian particles and ask how many particles of this ensemble have at time t the velocities in the interval $(v, v + dv)$ described by the probability density $P(v, t)$. The deterministic equation for $P(v, t)$ is called the Fokker-Planck , and for the white noise (2.4) has the following form

$$\frac{\partial P(v,t)}{\partial t} = \frac{\partial}{\partial v}\left[\gamma v P\right] + D\frac{\partial^2 P}{\partial v^2}. \tag{3.16}$$

Multiplying (3.16) by v and v^2 and performing the integrations by parts on the right hand side of this equation, one gets for the first two moments $\langle v \rangle$ and $\langle v^2 \rangle$,

$$\langle v \rangle = v(0)\exp\left(-\frac{\gamma}{m}t\right);$$

$$\langle v^2 \rangle = v(0)^2\exp\left(-\frac{2\gamma}{m}t\right) + \frac{D}{m\gamma}\left[1 - \exp\left(-\frac{2\gamma}{m}t\right)\right]. \tag{3.17}$$

Let us demonstrate [34] the equivalence of the Fokker-Planck equation

$$\frac{\partial P(x,t)}{\partial t} = -\frac{\partial}{\partial x}\left[f(x)P\right] + D\frac{\partial^2 P}{\partial x^2} \tag{3.18}$$

and the Langevin equation

$$\frac{dx}{dt} = f(x) + \xi(t), \tag{3.19}$$

with an arbitrary (nonlinear) function $f(x)$ and white noise $\xi(t)$.

Integrating Equation (3.19) between t and $t + \Delta t$ one gets

$$x(t+dt) - x(t) \equiv \Delta x = \int_t^{t+\Delta t} f[x(u)]\,du + \int_t^{t+\Delta t} \xi(u)\,du. \tag{3.20}$$

Averaging this equation over $\xi(t)$ at constant $x(t)$ gives

$$\langle \Delta x \rangle = f[x(t)]\Delta t + O\left[(\Delta t)^2\right] \tag{3.21}$$

which gives the first term in (3.18). Then,

$$\left\langle (\Delta x)^2 \right\rangle = \left\langle \left[\int_t^{t+\Delta t} f\left[x\left(u\right)\right] du \right]^2 \right\rangle + \tag{3.22}$$

$$2 \int_t^{t+\Delta t} du \int_t^{t+\Delta t} dz \left\langle f\left[x\left(u\right)\right] \xi\left(z\right) \right\rangle + \int_t^{t+\Delta t} du \int_t^{t+\Delta t} dz \left\langle \xi\left(u\right) \xi\left(z\right) \right\rangle.$$

The first two terms in (3.22) are of the order of $(\Delta t)^2$ and can be neglected, just as in (3.21) while the last term gives $D\Delta t$ which shows the equivalence of the Langevin Equation (3.19) and the Fokker-Planck Equation (3.18). One additional comment is required to show the full equivalence of these equations. The Fokker-Planck equation defines the distribution function $P(x)$ which allows one to calculate all moments of $x(t)$ while the Langevin equation contains only the first two moments. To eliminate this contradiction we will consider only those delta-correlated noises (the so-called Gaussian white noises) for which the first two moments define all higher moments, namely all odd moments vanish, and all even moments are

$$\left\langle \xi\left(t_1\right) \xi\left(t_2\right) ... \xi\left(t_n\right) \right\rangle = \sum \left(2D\right)^n \delta\left(t_1 - t_2\right) ... \delta\left(t_{n-1} - t_n\right). \tag{3.23}$$

The summation in (3.23) extends over all different ways in which the indices $1, 2, ...n$ can be subdivided in pairs.

3.2.2 *Multiplicative white noise*

The random force entered Equation (3.1) additively. The situation is more problematic for noise which appears multiplicatively in an equation of motion,

$$\frac{dx}{dt} = f\left[x\left(t\right)\right] + g\left[x\left(t\right)\right] \xi\left(t\right). \tag{3.24}$$

Usually the functions $f(x)$ and $g(x)$ are powers of x, $f \sim x^n$ and $g \sim x^m$. Then, Equation (3.24) takes the form

$$\frac{dx}{dt} = ax^n + bx^m \xi\left(t\right). \tag{3.25}$$

By introducing the new variable $y = x^{1-m}$ one can convert Equation (3.25) with multiplicative noise to the form

$$\frac{dy}{dt} = a\left(1 - m\right) y^{\frac{n-m}{1-m}} + b\left(1 - m\right) \xi\left(t\right) \tag{3.26}$$

with additive noise.

As it was discussed earlier after Equation (2.4), the real "white noise" $\xi(t)$ could be expressible as a sequence of randomly arriving delta peaks of finite width which causes a jump in $x(t)$. It is not clear, therefore, at which moment one has to take the value of x in the function $g[x(t)]$ (at the beginning, in the middle or at the end of a jump). The latter decision will lead to the different forms of the Fokker-Planck equation. The three possibility described above will lead to the Ito - Stratonovich dilemma [34] or kinetic [35] interpretation, respectively. Integrating Equation (3.24) one gets, analogously to (3.20),

$$x(t+\Delta t) - x(t) = f[x(t)]\Delta t + g[x(t)] \int_{t}^{t+\Delta t} \xi(u)\,du, \text{ (Ito)} \qquad (3.27)$$

$$x(t+\Delta t) - x(t) = f[x(t)]\Delta t + g\left[\frac{x(t)+x(t+\Delta t)}{2}\right] \int_{t}^{t+\Delta t} \xi(u)du, \text{ (Stratonovich)}$$

$$x(t+\Delta t) - x(t) = f[x(t)]\Delta t + g[x(t+\Delta t)] \int_{t}^{t+\Delta t} \xi(u)\,du. \text{ (kinetic)}$$

It follows from (3.27) [34] that the Fokker-Planck equation will be different in Ito and Stratonovich interpretation

$$\frac{\partial P(v,t)}{\partial t} = -\frac{\partial}{\partial x}\left\{\left[f(x) + \nu D g(x)\frac{dg}{dx}\right]P\right\} + D\frac{\partial^2}{\partial x^2}\left[g^2(x)P\right] \qquad (3.28)$$

with $\nu = 1$ (Stratonovich) and $\nu = 0$ (Ito).

Note that the Stratonovich interpretation is more natural since it follows from the usual calculus. Indeed, let us change the variables in Equation (3.24)

$$x_1 = \int \frac{dx}{g(x)}; \quad f(x_1) = \frac{f(x)}{g(x)}; \quad P(x_1) = P(x)g(x) \qquad (3.29)$$

which gives

$$\frac{\partial P(x_1,t)}{\partial t} = -\frac{\partial}{\partial x_1}[f(x_1 P)] + D\frac{\partial^2 P}{dx_1^2} \qquad (3.30)$$

and, after transforming back to the original variable x, one gets the Stratonovich form of the Fokker-Planck Equation (3.28) with $\nu = 1$.

Let us mention also an another way of getting the Fokker-Planck equation in the presence of both additive and multiplicative white noises [36]

$$\frac{dx}{dt} = f[x(t)] + g[x(t)]\xi(t) + \eta(t) \qquad (3.31)$$

where

$$\langle \xi(t_1)\xi(t_2)\rangle = 2D\delta(t_1 - t_2); \quad \langle \eta(t_1)\eta(t_2)\rangle = 2\alpha\delta(t_1 - t_2). \qquad (3.32)$$

Then, the Fokker-Planck Equation (3.16) transforms into a stochastic equation

$$\frac{\partial P(x,t,\xi)}{\partial t} = -\frac{\partial}{\partial x}\left[\{f(x) + g[x(t)]\xi(t)\}P\right] + \alpha\frac{\partial^2 P}{\partial x^2}. \qquad (3.33)$$

We are interested in the probability density $P(x,t)$, defined as the average of $P(x,t,\xi)$ over all realizations of $\xi(t)$. Averaging (3.33) one gets

$$\frac{\partial P(x,t)}{\partial t} = -\frac{\partial}{\partial x}\left[f(x)P\right] + \alpha\frac{\partial^2 P}{\partial x^2} - \frac{\partial}{\partial x}\left\{g[x(t)]\langle\xi(t)P(x,t,\xi)\rangle_\xi\right\}. \qquad (3.34)$$

The last term in (3.34) can be computed using the Novikov-Furutzu theorem for a Gaussian process [28]

$$\langle \xi(t)P(x,t,\xi)\rangle_\xi = \int_0^t dz\, \langle\xi(t)\xi(z)\rangle \left\langle \frac{\partial P(x,t,\xi)}{\partial \xi(z)}\right\rangle \qquad (3.35)$$

$$= -D\frac{\partial}{\partial x}\left\{g[x(t)]P(x,t)\right\}.$$

Inserting this result in (3.34) one gets

$$\frac{\partial P(x,t)}{\partial t} = -\frac{\partial}{\partial x}\left\{\left[f(x) + \alpha\frac{\partial g}{\partial x}\right]P(x,t)\right\} + \frac{\partial^2}{dx^2}(\{\alpha + Dg^2[x(t)]\}P(x,t)) \qquad (3.36)$$

which for $\alpha = 0$ coincides with (3.28).

3.3 Active Brownian particle

Thus far, we have considered the equilibrium state of a Brownian particle, where the energy is due to noise, and is transferred to a surrounding medium. The energy balance is described by the fluctuation-dissipation theorem. Such a description of equilibrium system dates back to Gibbs, and can be applied to all equilibrium systems. The situation is more complicated for non-equilibrium systems, where there is no universal approach, and one has to consider each class of non-equilibrium systems separately,

Here, we consider the special class of Brownian motion, the so-called "active motion", where a particle is able to gain kinetic energy from the environment and transfer it, under non-equilibrium conditions, to direct motion (self-propelling mechanism), instead of transferring the energy to the environment. Such processes are clearly very important in modelling biological processes. A detailed description is given in the extensive review with 385 references [37]. We shall restrict our considerations to the general idea of the method, which consists of introducing velocity-dependent friction. Then, the motion of Brownian particle (with $m = 1$) is described by the following equations,

$$\frac{d\mathbf{r}}{dt} = \mathbf{v}; \qquad \frac{d\mathbf{v}}{dt} = -\gamma(\mathbf{v})\,\mathbf{v} - \nabla V(r) + \boldsymbol{\xi}(t) \tag{3.37}$$

where $V(r)$ is an external potential, $\gamma(\mathbf{v})$ is the effective friction, and $\xi(t)$ is white noise of strength D, independent of friction since the fluctuation-dissipation theorem is invalid for non-equilibrium states. The different models are distinguished by the form of the function $\gamma(\mathbf{v})$. In the simplest Rayleigh-Helmholtz model, one assumes a parabolic shape for this function,

$$\gamma(v) = -\alpha + \beta v^2 = \beta\left(v^2 - v_0^2\right) \tag{3.38}$$

where $v_0^2 = \alpha/\beta$ defines a special value of velocity below which the friction becomes negative, and the particle starts to gain the energy from the surrounding medium. The stationary solution of the Fokker-Planck equation for the probability distribution $P(\mathbf{r}, \mathbf{v}, t)$, corresponding to the Langevin Equation (3.37)

$$\frac{\partial P}{\partial t} = -\mathbf{v}\frac{\partial P}{\partial r} - \nabla U(\mathbf{r})\frac{\partial P}{\partial \mathbf{v}} + \frac{\partial}{\partial \mathbf{v}}\left[\gamma(\mathbf{v})\,vP + D\frac{\partial P}{\partial \mathbf{v}}\right] \tag{3.39}$$

has the following form [38],

$$P_{st.}(\mathbf{v}) = N\exp\left[\frac{\beta\mathbf{v}^2}{2D}\left(v_0^2 - \frac{\mathbf{v}^2}{2}\right)\right]. \tag{3.40}$$

Equation (3.40) shows that at $v_0^2 = 0$ the noisy system bifurcates to a limit cycle, and above this bifurcation, the particles moves forever.

Another model is based on the function $\gamma(v) = \gamma_0(1 - v_0\,|\,v\,|^{-1})$, which leads to $P_{st.} = N\exp\left[-(\gamma_0/2D)(|\,v\,| - v_0)^2\right]$. More complicated models include the idea of "energy depot", where the particle is capable of storing some energy in an internal depot and afterwords converting part of this energy into kinetic energy of motion [37].

3.4   Color noise: unified color noise approximation (UCNA)

There is no rigorous way to go from the Langevin equation

$$\frac{dx}{dt} = f(x) + \xi_1(t)\,g_1(x) + \xi_2(t)\,g_2(x) \tag{3.41}$$

with color noises $\xi_{1,2}(t)$ to the appropriate Fokker-Planck equation. We consider the best known approximation, the UCNA [39] scheme for the Ornstein-Uhlenbeck noise with the correlator

$$\langle \xi_i(t_1) \xi_i(t_2) \rangle = D_i \exp\left(-\frac{|t_1 - t_2|}{\tau_i}\right), \quad i = 1, 2. \tag{3.42}$$

One can rewrite Equations (3.41)-(3.42) in the following form [40]

$$\frac{dx}{dt} = f(x) + \xi_1(t) g_1(x) + \xi_2(t) g_2(x), \tag{3.43}$$

$$\frac{d\xi_i}{dt} = -\frac{1}{\tau_i} \xi_i + \sqrt{D_i} \varepsilon_i(t)$$

where $\varepsilon(t)$ is white noise, $\langle \varepsilon_i(t_1) \varepsilon_i(t_2) \rangle = 2\delta(t_1 - t_2)$. On eliminating ξ_i from Equations (3.43) and introducing the new variable $s = \tau^{-\frac{1}{2}} t$, one gets

$$\frac{d^2x}{ds^2} + \eta(x, \tau) \frac{dx}{ds} - f(x) = \frac{\sqrt{D}}{\tau^{1/4}} \varepsilon(s), \tag{3.44}$$

where for additive noise the nonlinear damping $\eta(x, \tau)$ reads

$$\eta(x, \tau) = \frac{1}{\sqrt{\tau}} + \sqrt{\tau}\left(-\frac{df}{dx}\right), \tag{3.45}$$

and for multiplicative noise,

$$\eta(x, \tau) = \frac{1}{\sqrt{\tau}} + \sqrt{\tau}\left[-\frac{df}{dx} + f(x) \frac{d(\ln g_1(x))}{dx}\right]. \tag{3.46}$$

If the noise-free Equation (3.41) is stable, then $\frac{df}{dx} < 0$, and the expressions in the brackets in (3.45) and (3.46) are positive. Moreover, $\eta(x, \tau)$ approaches infinity for both $\tau \to 0$ and $\tau \to \infty$. Therefore, for large positive damping one can neglect the second derivatives in Equation (3.44), and rewrite this equation in terms of the original variable $t = \sqrt{\tau} s$ as

$$\eta(x, t) \frac{dx}{dt} = f(x) + [g_1(x) D_1 + g_2(x) D_2] \varepsilon(t) \tag{3.47}$$

with

$$\eta(x, t) = 1 - \sum_{i=1,2} \tau_i \left[\frac{df}{dx} - \frac{d \ln g_i}{dx_i} f(x)\right]. \tag{3.48}$$

For example, for the overdamped double-well oscillator with multiplicative and additive noise

$$\frac{dx}{dt} = ax - bx^3 + \xi_1(x) x + \xi_2(x) \tag{3.49}$$

one gets [41]

$$\eta = 1 - a\tau_2 + (2\tau_1 + 3\tau_2) bx^2.$$

Note, that Equation (3.47) becomes exact both at $\tau \to 0$ and $\tau \to \infty$, and hence is expected to be a good approximation for intermediate τ as well. In such a way Equation (3.41) with color noise is reduced to Equation (3.47) with white noise.

3.5   Brownian motion and anomalous diffusion

As we have seen in section 3.1, a Brownian particle subjected to white noise is characterized by the normal diffusion where for long times $\left\langle (\Delta x)^2 \right\rangle$ increases linearly with time. However, it turns out that in many fields of natural science (dozens of examples can be found in the comprehensive reviews [42], [43]) diffusion is anomalous, namely for large t

$$\left\langle (\Delta x)^2 \right\rangle \sim t^\alpha \qquad (3.50)$$

with $\alpha \neq 1$. The cases with $\alpha < 1$ and $\alpha > 1$ are called subdiffusion and superdiffusion, respectively.

We show now [44] that the anomalous diffusion can be obtained from the standard Langevin Equation (3.1) with the non-white correlator $r(z)$ defined in (2.1). On the substitution of (2.1) into (3.4), one gets for the dispersion of the particle position $\left\langle x^2(t) \right\rangle$, with $x(t = 0) = 0$,

$$\left\langle x^2(t) \right\rangle = \frac{1}{\gamma m} \int_0^t dz F(z) \left\{ 2 \exp\left[-\frac{\gamma}{m}(t - z) \right] \right.$$
$$\left. - \exp\left[-\frac{\gamma}{m}(2t - z) \right] - \exp\left(-\frac{\gamma}{m} z \right) \right\} \qquad (3.51)$$

where

$$F(z) = \int_0^{|z|} du\, r(u)\,(|z| - u). \qquad (3.52)$$

It turns out [44], that

a) For $0 < F(\infty) < \infty$, (which requires $\lim_{v \to \infty} \int_0^v r(u)\, du \sim 1/v$) $\left\langle x^2(t) \right\rangle$ tends to a finite limit as $t \to \infty$, i.e., the diffusion is confined (stochastic localization).

b) For $F(\infty) = \infty$, the type of diffusion depends on the noise strength $R = \int_0^\infty r(z)\, dz$, namely, $0 < R < \infty$ corresponds to normal diffusion, $R = \infty$ results in superdiffusion, and $R = 0$ (contribution from regions of positive and negative correlations balance) gives subdiffusion.

One can obtain the anomalous diffusion even in the case of white noise in Equation (3.1), if the viscosity coefficient η is time-dependent, $\eta = \eta(t)$, say having a form

$$\eta = \frac{a}{t^\beta}. \qquad (3.53)$$

It turns out [45], that for $0 < \beta < 1$ one gets the subdiffusion for $a > 0$ and the stretched exponent $\left\langle x^2 \right\rangle \approx \exp\left(-2|a|\left(t^{1-\beta} \right) / (1 - \beta) \right)$ for $a < 0$, $\left\langle x^2 \right\rangle > \approx t \ln t$ dependence for $\beta = 1$ and $a < 1/2$, and localization for $\beta < 0$ and $a > 0$.

3.6 Brownian motion in a fluctuating medium

Another generalization of the classical Brownian motion has to be performed when the Brownian particle is moving in a medium with great fluctuations. The latter can be the fluctuations of density, as near the critical point, or the fluctuations of energy, as in Josephson junction with a fluctuating shunt resistor. The first case we will consider in the next chapter within the special model, whereas here we consider the general case with the fluctuations of the damping coefficient in Equation (3.1) [46], which leads to the following equation of motion,

$$m\frac{dv}{dt} = -\gamma(z) + \sqrt{2\gamma(z)\kappa T}\xi(t), \qquad (3.54)$$

where $z(t)$ is some characteristic of a surrounding media and $\xi(t)$ is white noise, connected with the damping coefficient by the fluctuation-dissipation theorem. The simplest example of the influence of the surrounding media is the free diffusion of a Brownian particle with two possible values of viscosity (non-symmetric dichotomous noise with the correlation time τ)

$$\gamma = \gamma_o + z(t), \qquad (3.55)$$

where $z(t)$ takes either value $-a$ or b with $\langle z \rangle = 0$ and $0 \leq a \leq \gamma_0$, $b \geq 0$. For this special case, the extended Langevin Equation (3.54) takes the following form,

$$m\frac{dv}{dt} + (\gamma_0 - a)v = \sqrt{2(\gamma_0 - a)\kappa T}\xi(t) \qquad (3.56)$$

$$m\frac{dv}{dt} + (\gamma_0 + b)v = \sqrt{2(\gamma_0 + b)\kappa T}\xi(t)$$

with $dx/dt = v$. Multiplying these equations by v and x, one obtains, after averaging, five coupled equations for the second moments of x and v. It turns out [46] that $lim_{t\to\infty}\langle x(t)v(t)\rangle$ is finite, i.e. the particle's motion is diffusive, and the diffusion coefficient D is defined by this finite value. The simple calculation leads to

$$D = D_0\left[1 + \frac{ab}{(\gamma_0 - a)(\gamma_0 + b) + m\gamma_0/\tau}\right] \qquad (3.57)$$

where $D_0 = \kappa T/\gamma_0$ is the diffusion coefficient of usual Brownian motion with the friction coefficient γ_0. Clearly, $D > D_0$. However, the diffusion coefficient D approaches its classical value D_0 when $\tau \to 0$ or $ab \to 0$. The former case corresponds to fast fluctuations of damping coefficients, when, due to its inertia, the Brownian particle does not feel the fluctuations, while the later case corresponds to weak fluctuations.

3.7   Brownian motion near the critical point

The foregoing theory of Brownian motion was based on the small parameter $l/L < 1$, where L is the characteristic size of the Brownian particle, and l is the size of the surrounding molecules. One can consider the opposite limit case, $l/L > 1$, where l is the characteristic size of an inhomogeneity in the medium, such as the size of clusters near the liquid-gas critical point or smallest scale of turbulence in a turbulent flow. The cause of stochasticity is completely different in these two cases. While in the usual Brownian motion the random force is associated with random collisions of molecules with the Brownian particle, in the latter case the stochasticity of the motion of the particle, which we still call the "Brownian particle", is provided by the finite lifetime of clusters and randomness of their decay. The Brownian particle is captured by a cluster, moves for some time inside the given cluster, and after the decay of this cluster the particle is captured by another cluster and so on. For simplicity, we assume [47] that immediately after release by a cluster the particle is captured by another cluster, so in the frame moving with the mean velocity (if such velocity exists) the equation of motion of the particle inside a cluster has the following form

$$\frac{dv}{dt} = -\gamma_0 (v - w) \tag{3.58}$$

where $\gamma_0 = \gamma/m$, and v and w are the velocities of a particle and a cluster, respectively. It is assumed in (3.58) that the drag force acting upon the particle is linear in its velocity relative to that of the cluster at the same point and time. Equation (3.58) has the familiar form (3.1) describing Brownian motion with a random force $\xi(t) = \gamma_0 w$. However, if this noise is not white one has to rewrite Equation (3.58) in the form

$$\frac{dv}{dt} - \int_0^t \gamma(t - z) v(z) \, dz = \xi(t) \tag{3.59}$$

and according to the so-called second fluctuation - dissipation theorem [48],

$$\langle \xi(t)\,\xi(0) \rangle = \langle v^2 \rangle \, \gamma(t). \tag{3.60}$$

In the case of white noise $\gamma(t) = \gamma_0 \delta(t)$, and (3.59) reverts to (3.58). We would like to compare the diffusion coefficients D of a Brownian particle and ε of a medium. These coefficients can be expressed as integrals of the velocity correlation functions [48],

$$D = \int_0^\infty \langle v(0)\,v(z) \rangle \, dz; \tag{3.61}$$

$$\varepsilon = \int_0^\infty \langle w(0)\,w(z) \rangle \, dz = \frac{1}{\gamma_0^2} \int_0^\infty \langle \xi(0)\,\xi(z) \rangle \, dz.$$

On multiplying Equation (3.59) by $v(0)$, averaging the resulting equation, and taking into account the statistical independence of the velocities of particle and medium, $\langle v(0) w(t) \rangle = 0$, one obtains a solution for the correlation function of a Brownian particle $y(t) \equiv \langle v(t) v(0) \rangle$,

$$y(t) = \langle v^2 \rangle - \int_0^t d\tau \int_0^\tau \gamma(t - z) y(z) dz = \langle v^2 \rangle - \int_0^t y(s) ds \int_s^t \gamma(\tau - s) d\tau.$$
(3.62)

The order of integration was changed on passing to the last equality in (3.62). On substituting Equation (3.60) into (3.62), taking $t \to \infty$, with $y(\infty) = 0$, one derives from (3.62) that

$$\int_0^\infty y(s) ds \int_0^\infty \langle w(0) w(z) \rangle dz = \langle v^2 \rangle^2.$$
(3.63)

Finally, by using (3.61), we obtain from (3.63) that

$$D = \frac{\langle v^2 \rangle^2}{\gamma_0^2 \epsilon}.$$
(3.64)

One can rewrite Equation (3.64) using the fluctuation-dissipation theorem (3.60) with $\xi = \gamma_0 w$. Then, the denominator in Equation (3.64) is reduced to $\langle v^2 \rangle \int_0^\infty \gamma(t) dt = \langle v^2 \rangle \gamma(0)$, where $\gamma(0)$ is the Fourier -component of the function $\gamma(t)$ with zero frequency. Finally, Equation (3.64) reduces to

$$D = \frac{\langle v^2 \rangle}{\gamma(0)}$$
(3.65)

which is formally similar to the Einstein formula for the diffusion coefficient $D = \kappa T / \gamma = \langle v^2 \rangle / \gamma$.

All our previous analysis was based on the fluctuation-dissipation theorem which presumed the interaction and energy exchange between a Brownian particle and surrounding medium leading to the equilibrium. A similar situation takes place in particle diffusion in plasma turbulence, where this interaction is governed by electromagnetic interaction [49]. However, the situation is different in the case of diffusion of a small particle in a turbulent flow, since the fluctuation-dissipation theorem does not apply in this case [50]. As a result [50], in this case $D = \epsilon$.

The problem considered in this section is a special case of the general theory of the two-state random walk in which a particle can be in one of two states for a random period of time, with each of the states having different dynamics and a different switch density for a jump to the second state [51].

Chapter 4

Overdamped Harmonic Oscillator with Additive Noise

If one neglects the inertia term in the dynamic equation of a nonlinear oscillator with additive random force $\xi(t)$, one gets the stochastic differential equation of the form

$$\frac{dx}{dt} + f(x) = \xi(t) \tag{4.1}$$

or the linearized equation

$$\frac{dx}{dt} + ax = \xi(t). \tag{4.2}$$

In the case of white noise, one can replace the nonlinear Langevin equation (4.1) by the appropriate Fokker-Planck equation as it was described in section 3.2, while for the color noise one has to use some approximations such as UCNA considered in section 3.4. On the other hand, the linear equation (4.2) can be solved for an arbitrary form of noise. In the absence of random forces, the solution of equation (4.2) has a form $x(t) = x(0)\exp(-at)$, i.e., it vanishes at $t \to \infty$. On the other hand, in the absence of viscosity, $a = 0$, the second moment $\langle x^2 \rangle$ diverges in time. The solution in this case is [44],

$$\langle x^2(t) \rangle = 2F(t) \tag{4.3}$$

where $F(u)$ is defined in (3.52). If $F(\infty) < \infty$, $\langle x^2(t) \rangle$ tends to a finite value at $t \to \infty$. For $F(\infty) = \infty$, $\langle x^2(t) \rangle$ diverges at $t \to \infty$ (stochastic acceleration of the particle by random driving force). For white noise $\xi(t)$, $F(u) = Du$ (Fermi acceleration [52]).

Coming now back to equation (4.2), one can easily obtain the solution of this equation, with zero initial condition $x(0) = 0$,

$$\langle x^2(t) \rangle = \int_0^t dt_1 \int_0^t dt_2 \exp\left[-a(2t - t_1 - t_2)\right] \langle \xi(t_1)\xi(t_2) \rangle \tag{4.4}$$

which after substitution of (2.1) and simple transformations takes the form (3.51)

$$\left\langle x^2\left(t\right)\right\rangle = a \int_0^t duF\left(u\right)\left[\exp\left(-au\right) - \exp[-a\left(2t - u\right)] + 2F\left(t\right)\exp\left(-at\right).$$

$$(4.5)$$

In general, the correlator between $\xi\left(t\right)$ and $\xi\left(t_1\right)$ decreases as $|t - t_1|$ increases. Then, the asymptotic behavior of (4.5) is defined by the Laplace transform of $F\left(u\right)$,

$$\left\langle x^2\left(t\right)\right\rangle_{t\to\infty} = a\int_0^\infty duF\left(u\right)\exp\left(-au\right).\qquad(4.6)$$

4.1   Additive white noise

For additive white noise $\xi\left(t\right)$, the Fokker-Planck equation (3.16) associated with the Langevin equation (4.2) has the following form

$$\frac{\partial P}{\partial t} = \frac{\partial}{\partial x}\left(axP\right) + D\frac{\partial^2 P}{\partial x^2}.\qquad(4.7)$$

This equation was solved earlier in (3.16)-(3.17) leading to

$$\left\langle x\right\rangle = x\left(0\right)\exp\left(-at\right),\quad \left\langle x^2\right\rangle = x\left(0\right)^2\exp\left(-2at\right) + \frac{D}{a}\left[1 - \exp\left(-2at\right)\right].$$

$$(4.8)$$

4.2   Additive noise and periodic force

A solution of equation

$$\frac{dx}{dt} + ax = A\cos\left(\Omega t\right) + \eta\left(t\right)\qquad(4.9)$$

can be obtained analogously to (4.7)-(4.8) by replacing ax in (4.7) by $ax - A\cos\left(\Omega t\right)$. Then the final results will contain the oscillating function which performs a periodic motion around $x\left(0\right)$ in the presence of the periodic force,

$$\left\langle x\right\rangle = x\left(0\right)\exp\left(-at\right) + \frac{A}{a^2 + \Omega^2}\left\{a\left[\left(\cos\left(\Omega t\right) - 1\right)\right] + \Omega\sin\left(\Omega t\right)\right\}.\quad(4.10)$$

Chapter 5

Overdamped Harmonic Oscillator with Multiplicative Noise

5.1 Multiplicative noise (shift of stable points)

Let us start with the simplest form of multiplicative noise.

$$\frac{dx}{dt} = -ax + \xi(t)x. \tag{5.1}$$

As we have already seen, in the absence of noise, the solution of this equation vanishes at $t \to \infty$. The solution of (5.1) is

$$x(t) = \exp\left(-at + \int_0^t \xi(z)\,dz\right) \tag{5.2}$$

and the first moment will have the following forms depending of the type of noise [53]: for white noise,

$$\langle x(t) \rangle = x(0)\exp\left[(-a + D)t\right], \tag{5.3}$$

and for dichotomous noise,

$$\langle x(t) \rangle = x(0)\exp(-at)\frac{b_1\exp(b_2 t) - b_2\exp(b_1 t)}{b_1 - b_2}, \tag{5.4}$$

$$b_{1,2} = -\lambda \pm \sqrt{\lambda^2 + \sigma^2}, \quad \sigma^2 = (D/a)[1 - \exp(-2at)].$$

One sees from (5.3)-(5.4), that, by virtue of multiplicative noise, the first moment is not only not zero, but diverges at $t \to \infty$ for $D > a$, and $\sigma^2 > a^2 + 2a\lambda$, respectively.

From (5.2) one gets for the first moment

$$\langle x(t) \rangle = x(0)\exp\left(-at + \frac{1}{t}\int_0^t \int_0^t \langle \xi(t_1)\xi(t_2)\rangle \, dt_1 dt_2\right). \tag{5.5}$$

In the limit $t \to \infty$ this equation takes the form

$$\langle x(t) \rangle = x(0)\exp\left[-at + \pi\overline{S}(0)\right] \tag{5.6}$$

where

$$\overline{S}(0) = \lim_{t \to \infty} \frac{1}{t} \int_0^t S(z,0)\, dz \qquad (5.7)$$

is the zero component of the Fourier transform of the correlator (power spectrum),

$$S(t,\omega) = \frac{1}{2\pi} \int_{-\infty}^{\infty} \langle \xi(t+z)\xi(t) \rangle \exp(i\omega z)\, dz. \qquad (5.8)$$

These formulas show that the effect of multiplicative (parametric) noise at large times is due to the low-frequency (close to zero) noise power spectrum. This is exactly the opposite of that in the case of the parametric periodic force [54]. Indeed, the stabilization of the upper vertical state of the pendulum is achieved by the high-frequency harmonic vibrations of its suspension [55].

5.2 Multiplicative and additive noises

The simplest generalization of (5.1) involves, along with a multiplicative noise, an additional additive noise,

$$\frac{dx}{dt} = -ax + \sqrt{D}\xi(t)x + \sqrt{\alpha}\eta(t). \qquad (5.9)$$

Equation (5.9) has been intensively studied in different problems. To mention just a few: a quantum dimer [56], a coupled neutron network [57], synchronization of chaotic oscillators [58], detecting the gravitational background [59], and medical imaging [60].

5.2.1 *Two white noises*

Under the simplest assumption that two sources of noise in (5.9) are uncorrelated having the equal strength,

$$\langle \xi(t_1)\xi(t_2) \rangle = \langle \eta(t_1)\eta(t_2) \rangle = 2\delta(t_1 - t_2), \qquad (5.10)$$

one can get the exact solution [61] even for the case when the coefficient a is the arbitrary function of time, $a = a(t)$.

The solution of Equation (5.9) is given by

$$x(t) = K(t)L(t)\left[x(t=0) + \sqrt{\alpha} \int_0^t \frac{\eta(t_1)\, dt_1}{K(t_1)L(t_1)} \right] \qquad (5.11)$$

where

$$K(t) = \exp\left[-\int_0^t a(t_1)\,dt_1\right]; \quad L(t) = \exp\left[\sqrt{D}\int_0^t \xi(t_1)\,dt_1\right]. \quad (5.12)$$

For $a = \text{const.}$, the first two average moments are

$$\langle x(t)\rangle = x(t=0)\exp\left[-(a-D)t\right] \quad (5.13)$$

$$\langle x^2(t)\rangle = x^2(t=0)\exp\left[-2(a-2D)t\right] + \frac{\alpha}{a-2D}\left\{1 - \exp\left[-2(a-2D)t\right]\right\}.$$

Two limit cases can be obtained from (5.13). For $\alpha = 0$ and $a = \text{const.}$, one gets the moments of a system without additive noise, and for $D = 0$, those without multiplicative noise.

5.2.2 Two correlated white noises

Let us consider Equation (5.9) with $a = \text{const.}$ and two sources of white noise (5.10) which are correlated,

$$\langle \xi(t_1)\eta(t_2)\rangle = 2\kappa\sqrt{D\alpha}\,\delta(t_1 - t_2). \quad (5.14)$$

Correlations between different noise sources may occur when they both have the same origin, as in laser dynamics [62], or when strong noise leads to an appreciate change in the internal structure of a system and hence in its internal noise. There are few simple rules for obtaining the solution of a problem with correlated multiplicative and additive noise in terms of the results with non-correlated noise [63].

It turns out that for delta-correlated noise (5.14),

$$\langle x(t)\rangle = x(t=0)\exp\left[-(a-D)t\right] + \frac{\kappa\sqrt{D\alpha}}{a-D}\left\{1 - \exp\left[-(a-D)t\right]\right\};$$

$$(5.15)$$

$$\langle x^2(t)\rangle = \frac{\alpha}{a-2D} + \left[x^2(t=0) - \frac{\alpha}{a-2D}\right]\exp\left[-2(a-2D)t\right] +$$

$$6\kappa\sqrt{D\alpha}\int_0^t \langle x(u)\rangle \exp\left[-2(a-2D)u\right]du$$

which for $\kappa = 0$ reduces to (5.13). The stationary ($t \to \infty$) correlation function is given by

$$\langle x(t)x(t+\tau)\rangle_{t\to\infty} = \frac{\alpha}{a-2D} + \frac{4\kappa^2 D\alpha}{(a-D)^2} +$$

$$\frac{4\kappa^2\alpha D(2a-D)}{(a-D)^2(a-2D)}\exp\left[-(a-D)|\tau|\right] \quad (5.16)$$

5.2.3　*Two correlated dichotomous noises*

The detailed analysis of correlated multiplicative and additive noises in linear systems can be found in [64], [65]. Here we consider the influence on stochastic resonance, which will be considered in the next section, of correlations between two multiplicative dichotomous noise. Such a system is described by the following equation

$$\frac{dx}{dt} = gx + c\xi_1(t)x + d\xi_2(t)x + A\sin(\omega t) + r \qquad (5.17)$$

where

$$\langle \xi_i(t_1)\xi_i(t_1)\rangle = D_i\lambda_i\exp\left(-\lambda_i|t_1 - t_2|\right); \qquad i = 1,\,2 \qquad (5.18)$$
$$\langle \xi_1(t_1)\xi_2(t_1)\rangle = \kappa\sqrt{D_1D_2\lambda_1\lambda_2}\exp\left(-\sqrt{\lambda_1\lambda_2}\,|t_1 - t_2|\right).$$

In the absence of correlations between $\xi_1(t)$ and $\xi_2(t)$, $(\kappa = 0)$, the presence of the second noise leads to an appearance of two stochastic resonances, when the amplitude of the periodic solution for $< x >$ has two maxima as the functions of the strengths of two noises [66]. The phenomenon of stochastic resonance shows only one maximum, if one of noises is Gaussian or Poisson white noise and disappears for two white noises. Since the intrinsic frequency of a system depends not only on D_1 and D_2, but also on κ, the correlation between noises can affect the stochastic resonance phenomenon. It turns out [67] that the value of κ affects the appearance of the stochastic resonance, which shifts the position of maximum and reaches the maximum value of resonance at some value of κ.

The influence of the correlation between additive and multiplicative noises, described by equation

$$\frac{dx}{dt} = 1 - bx + A\sin(\omega t) + \xi_1(t)x + \xi_2(t) \qquad (5.19)$$

with $\xi_1(t)$ and $\xi_2(t)$ defined in (5.18) was studied recently both analytically and numerically [65]. The output signal amplitude and SNR were found for different multiplicative noise intensity, signal frequency, noise symmetric parameter and noise correlation strength. It turns out that the output signal amplitude exhibits non-monotonic dependence on the multiplicative noise intensity, while shows monotonic decrease with signal frequency, and is independent of noise correlation strength. The SNR in the case of negative correlation is greater than those for uncorrelated and positively correlated noises.

5.3 Multiplicative color noise and periodic signal (stochastic resonance (SR) in linear systems)

The interesting phenomenon of stochastic resonance may appear in a system subject to both random and periodic force. Manifestation of SR is typically related to $x(t)$ at the characteristic frequency and can be expressed in terms of different functions that are calculated in terms of $x(t)$ such as the autocorrelation function, the power spectrum, or a signal-to-noise ratio which behave non-monotonically as a function of the noise amplitude, or the correlation rate.

It first seemed [68] that all three ingredients — non-linearity, periodic and random forces — are necessary for the onset of SR. However, it later became clear that SR may appear without a random force (replaced by a chaotic signal [69]), without a periodic force (autonomous SR [70], aperiodic SR [71]) or by replacing the characteristic frequency by some fluctuation rate [72]), and in linear systems (with multiplicative noise [73],[74]). The latter has been shown for the first-order differential equation describing an overdamped oscillator with multiplicative noise (Ornstein-Zernike [74], Gaussian [75], Poisson [76], or composite [77] noise).

Recently, not only the number of publications on stochastic resonance phenomenon is growing steadily, but also much extension of the conception of SR has appeared, such as doubly SR [78], stochastic giant-resonance [79], stochastic multiresonance [80], coherence resonance (generated in the system without an external force) [81], quantum SR [82], control of SR [83], autonomous SR [84], and aperiodic SR [85], etc.

Since the nonlinearity presents difficulties for the theoretical analysis, the linear models with multiplicative noise are of special interest. These models on the one hand show quasi-nonlinear behavior including stochastic resonance [73],[74], and on the other hand, allow an exact analytical treatment.

On addition of a periodic force to Equation (5.1) one gets

$$\frac{dx}{dt} = -ax + \xi(t)x + A\sin(\Omega t). \qquad (5.20)$$

The averaged solution of (5.20) has the following form

$$\langle x(t) \rangle = \exp(-at)\{x(0)\left\langle \exp\left(\int_0^t \xi(z)\,dz\right)\right\rangle +$$

$$A\int_0^t \sin(\Omega u)\exp(au)\left\langle \exp\left[\int_0^{t-u}\xi(z)\,dz\right]\right\rangle\}. \qquad (5.21)$$

Differentiating $\left\langle \exp\left(\int_0^t \xi(z)\, dz \right) \right\rangle$ one gets

$$\frac{d}{dt}\left\langle \exp\left(\int_0^t \xi(z)\, dz \right) \right\rangle = \left\langle \xi(t) \exp\left[\int_0^t \xi(z)\, dz \right] \right\rangle =$$

$$\left\langle \xi(t) \left[1 + \int_0^t \xi(z)\, dz + \ldots \right] \right\rangle = \int_0^t \left\langle \xi(t)\xi(u) \exp\left(-\int_0^u \xi(z)\, dz \right) \right\rangle du +$$

$$\int_0^t \langle \xi(t)\xi(u) \rangle \left\langle \exp\left[-\int_0^u \xi(z)\, dz \right] \right\rangle du. \qquad (5.22)$$

The splitting of average of product of three random forces in the last line in (5.22) is exact for white and dichotomous noises, and it is a quite good approximation for other types of color noise.

Using the Laplace transform to solve Equation (5.22), and substituting this solution into (5.21), one gets

$$\langle x(t) \rangle = \left[x(0) + \frac{A\Omega}{(a - b_1)^2 + \Omega^2} \right] \left(\frac{b_1}{b_1 - b_2} \right) \exp\left[(b_2 - a)\, t \right] +$$

$$\left[x(0) + \frac{A\Omega}{(a - b_2)^2 + \Omega^2} \right] \left(\frac{b_2}{b_2 - b_1} \right) \exp\left[(b_1 - a)\, t \right] +$$

$$A \left[\frac{\Omega^2 + (a + \lambda)^2}{\left[(a - b_1)^2 + \Omega^2 \right] \left[(a - b_2)^2 + \Omega^2 \right]} \right]^{1/2} \cos(\Omega t + \phi) \qquad (5.23)$$

where $b_{1,2}$ were defined in (5.4).

For white noise Equation (5.23) reduces to

$$\langle x(t) \rangle > = \left[x(0) + \frac{A\Omega}{(a - D)^2 + \Omega^2} \right]_1 \exp\left[-(a - D)\, t \right] +$$

$$\frac{A}{\left[(a - D)^2 + \Omega^2 \right]^{1/2}} \cos(\Omega t + \phi). \qquad (5.24)$$

For all $a > D$, where $\langle x(t) \rangle$ remains bounded, the amplitude of the stationary solution in Equation (5.24) is a monotonic function of D reaching maximum at $D = a$ showing some SR-like behavior [86].

The average stationary solution of Equation (5.20) can be obtained from the limit of (5.23) at $t \to \infty$ which gives

$$\langle x(t) \rangle_{st} = \frac{A\sqrt{\Omega^2 + (a + \lambda)^2}\, \cos(\Omega t + \phi)}{\sqrt{\Omega^4 + 2\left(a^2 + 2a\lambda + 2\lambda^2 + \sigma^2 \right)\Omega^2 + \left(a^2 + 2a\lambda - \sigma^2 \right)^2}}. \qquad (5.25)$$

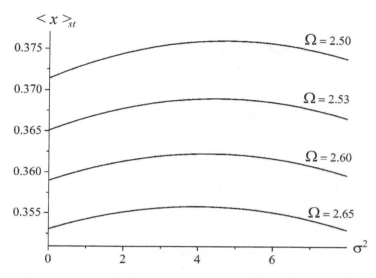

Fig. 5.1   The amplitude of a stationary signal as a function of the noise strength for a $= A = 1; \lambda = 10$ and different frequencies of the periodic fields.(Reprinted from Ref. 74 with permission from EDP Science).

The amplitude of the stationary solution depends on the dynamic parameter a, the amplitude A and the frequency Ω of the periodic force and the strength σ^2 and the correlation rate λ of the noise. Some typical graphs displayed in Figures 5.1 and 5.2 show the non-monotonic dependence of $\langle x(t) \rangle_{st}$ as a function of σ^2 and λ (stochastic resonance). Analysis of (5.25) shows that the maxima occurs at

$$\sigma^2_{max} = a^2 + 2a\lambda - \Omega^2. \tag{5.26}$$

The appearance of SR for a process described by Equation (5.20) with dichotomous noise $\xi(t)$ can be understood in the following way. For the limiting case $a = \xi = 0$, a particle executes a periodic motion with an amplitude A/Ω. If there is no random force, $\xi = 0$, but $a \neq 0$, a particle will move along the parabola $U = ax^2/2$. For dichotomous noise $\pm\sigma$, a particle is moving along the parabola $U = (a + \sigma)x^2/2$, then jumps at rate τ to the parabola $U = (a - \sigma)x^2/2$, etc. For $\sigma > a$, but $\sigma < (a^2 + 2a\lambda)^{1/2}$, these two parabolas have curvatures of opposite sign, and they act in opposite directions tending to increase (decrease) the displacement x of a particle. Their mutual influence is defined by noise which causes jumps between the parabolas and by a periodic force which determines the amplitude of

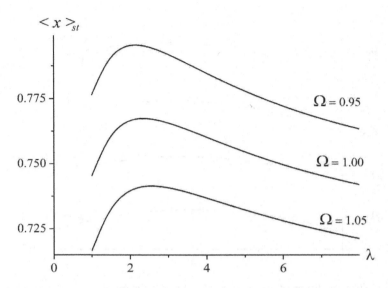

Fig. 5.2   The amplitude of a stationary signal as a function of the correlation time for $a = A = \sigma^2 = 1$ and different frequencies of the periodic field, [Reprinted from Ref. 74 with permission from EDP Sciences].

oscillations along the parabolas. Accordingly, the amplitude of the stationary signal, Eq. (5.25), has a maximum as a function of the noise strength.

Contrary to the dichotomous noise, the white noise has no characteristic frequency so there is no "adjustment" for the frequency Ω of an external field, and there is no stochastic resonance for white noise.

The more general case of additive and multiplicative color noise as well as a periodic signal has been considered in [64].

5.4   Stochastic resonance in a overdamped system with signal-modulated noise

In the previous section we considered an overdamped oscillator subject to multiplicative noise and an additive periodic signal. However, one can also consider the case when the signal is multiplied by noise. Such problem has been considered in [87] for bistable potentials in connection with the periodically modulated noise arising in some optical and astronomical devices. A similar problem has been considered for linear system [88]. We bring here the results of the latter article.

The overdamped oscillator is subject to a signal-modulated noise $\xi(t)$ and a non-correlated additive noise $\eta(t)$,

$$\frac{dx}{dt} + ax = \xi(t)[1 + A\cos(\Omega t)] + \eta(t) \qquad (5.27)$$

where $\xi(t)$ and $\eta(t)$ are exponentially correlated color noises

$$\langle \xi(t) \rangle = \langle \eta(t) \rangle = 0; \quad \langle \xi(t_1)\xi(t_2) \rangle = \frac{Q}{\tau} \exp\left(-\frac{|t_1 - t_2|}{\tau}\right); \qquad (5.28)$$

$$\langle \eta(t_1)\eta(t_2) \rangle = \frac{D}{\tau} \exp\left(-\frac{|t_1 - t_2|}{\tau}\right);$$

$$\langle \xi(t_1)\eta(t_2) \rangle = 2\kappa \frac{\sqrt{DQ}}{\tau} \exp\left(-\frac{|t_1 - t_2|}{\tau}\right).$$

The solution of Equation (5.27) has the following form

$$x(t) = \exp(-at)\left\{ x(t=0) + \int_0^t du \exp(au)\xi(u)[1 + A\cos(\Omega u)] + \right.$$

$$\left. \int_0^t du \exp(au)\eta(u) \right\}. \qquad (5.29)$$

From Equation (5.29) one gets for the stationary value of the power spectrum of the correlation function,

$$S(\omega) = \int_{-\infty}^{\infty} du \exp(i\omega u) \langle x(t+u)x(t) \rangle_{t\to\infty} = S_{\text{signal}} + S_{\text{noise}} \qquad (5.30)$$

with the signal-to-noise ratio ,

$$SNR = \frac{\int_{-\infty}^{\infty} S_{\text{signal}}(\omega) \, d\omega}{\int_{-\infty}^{\infty} S_{\text{noise}}(\omega - \Omega) \, d\omega}. \qquad (5.31)$$

The slightly cumbersome result of this exact calculation is [88]

$$SNR = \frac{aQA^2 a_+}{Q + D + 2\kappa\sqrt{QD}} \left[\frac{a_+ a_- + \Omega a_+}{2\left(a_-^2 + \Omega^2\right)\left(a_+^2 + \Omega^2\right)} + \right.$$

$$\left. \frac{a_+}{4a\left(a_+^2 + \Omega^2\right)} - \frac{a_-}{4a\left(a_-^2 + \Omega^2\right)} \right]$$

with $a_\pm = a \pm \tau^{-1}$.

Analysis of Equation (5.32) shows the existence of stochastic resonance: for the negative correlation region $-1 \leq \kappa < 0$, one finds a non-monotonic dependence of the signal-to-noise ratio SNR on both modulated noise strength Q and non-modulated noise strength D. It is remarkable that both effects remains even in the white noise limit, $\tau \to 0$. In addition, one gets a non-monotonic dependence of SNR on both the correlation rate τ^{-1} and the frequency Ω of the modulated field.

Chapter 6

Overdamped Single-well Oscillator

For the sake of simplicity, we start from the simplest non-linear oscillator model described by the following equation,

$$\frac{dx}{dt} = ax - bx^2 \tag{6.1}$$

with growth rate a and decay rate b.

The solution of Equation (6.1) is

$$x(t) = \frac{ax(t=0)}{bx(t=0) - [bx(t=0) - a]\exp(-at)} \tag{6.2}$$

which shows that the stability is defined by the initial position $x(t=0)$: the motion is stable for $x(t=0) > 0$, and it is unstable ($x(t) \to \infty$) for $x(t=0) < 0$.

For obvious reasons Equation (6.1) is called the birth-death equation for the growth of population, cells, etc., where x is assumed to be non-negative, $x \geq 0$. This equation has two stable points, $x_1 = 0$ and $x_2 = a/b$. The stability of these solutions is defined by linearizing Equation (6.1) around the stable points. This simple analysis shows that $x_1 = 0$ is stable for $a < 0$, whereas $x_2 = a/b$ is stable for $a > 0$. Let us now include the additive noise into Equation (6.1), and the fluctuations of parameters a and b leading to multiplicative noise, starting with the fluctuations of the parameter a.

6.1 Steady state

6.1.1 *White noises*

Introducing correlated white noise in (6.1), yields [89]

$$\frac{dx}{dt} = ax + x\xi(t) - bx^2 + \eta(t) \tag{6.3}$$

with

$$\langle \xi (t_1) \xi (t_2) \rangle = 2D\delta (t_1 - t_2) ; \quad \langle \eta (t_1) \eta (t_2) \rangle = 2\alpha\delta (t_1 - t_2) ; \quad (6.4)$$
$$\langle \xi (t_1) \eta (t_2) \rangle = 2\kappa\sqrt{D\alpha}\delta (t_1 - t_2) .$$

For a system subject only to multiplicative noise ($\eta = 0$ in (6.3)), there are [90] two critical values of the control parameter: $a = 0$ and $a = D$. For $a < 0$, the fixed point $x = 0$ is stable, and the stationary distribution function $P_{st} (x)$ is concentrated at zero as a delta-function, $P_{st} (x) \sim \delta (x)$. For $a > 0$, the fixed point $x = 0$ becomes unstable, but remains the most probable for $0 < a < D$. Finally, for $a > D$, the divergence at zero disappears and the point $x = a/b$ becomes stable.

The Fokker-Planck equation corresponding to the Langevin Equation (6.3) with noise (6.4) has the form [89]

$$\frac{\partial P}{\partial t} = -\frac{\partial}{\partial x} [A (x) P] + \frac{\partial^2}{\partial x^2} [B (x) P] \quad (6.5)$$

where the drift term $A (x)$ and the diffusive term $B (x)$ are

$$A (x) = ax - bx^2 + Dx - \kappa\sqrt{D\alpha}; \quad (6.6)$$
$$B (x) = Dx^2 - 2\kappa\sqrt{D\alpha}x + \alpha.$$

The time-dependent Equation (6.5) is quite complicated, but its stationary ($t \to \infty$) solution can be easily found. For $0 \leq \kappa < 1$, one gets

$$P_{st} (x) = NB (x)^{(a/2D)-1/2-\left(\alpha b^2 \kappa^2 / D^3\right)^{1/2}} \times \quad (6.7)$$
$$\exp \left[-\frac{bx}{D} + \frac{E}{\sqrt{D\alpha (1 - \kappa^2)}} \arctan \left(\frac{Dx - \kappa\sqrt{D\alpha}}{\sqrt{D\alpha (1 - \kappa^2)}} \right) \right]$$

where N is the normalization factor, and

$$E = \frac{b\alpha}{D} - \left[\kappa\sqrt{\frac{\alpha}{D}} \left(a + 2\kappa\sqrt{\frac{\alpha b^2}{D}} \right) \right] . \quad (6.8)$$

The question arises of the shift of the deterministic fixed points $x_1 = 0$, and $x_2 = a/b$ due to the presence of noise. As follows from (6.7), the extrema of $P_{st} (x)$ obey the following equation

$$bx^2 + (D - a) x - \kappa\sqrt{D\alpha} = 0. \quad (6.9)$$

The solutions of this equation for small correlation between noises κ are

$$x_1 \approx \frac{\kappa\sqrt{D\alpha}}{D - a}; \quad x_2 \approx \frac{a}{b} - \frac{D}{b} - \frac{\kappa\sqrt{D\alpha}}{D - a}. \quad (6.10)$$

Analysis of the influence of noise on $P_{st}(x)$ yields the following results [89]:

1. As the value of correlation parameter κ increases, $P_{st}(x)$ increases at small x and decreases at large x.

2. As the strength of the additive noise α increases, the maximum value of a peak increases at small value of x and decreases at large value of x. The peak is flattened and almost disappears at large values of α (effect of diffusion).

3. As the strength of the multiplicative noise D increases, the maximum of $P_{st}(x)$ moves to smaller values of x (effect of the drift term).

6.1.2 *Multiplicative noise (gene selection)*

A slightly different dimensionless form of Equation (6.3) has been used [91] to describe gene selection, namely,

$$\frac{dx}{dt} = a_1 - a_2 x + a_3 x (1 - x) + \xi(t) x (1 - x). \tag{6.11}$$

Special values of the parameters a_1 and a_2 have been chosen in the standard gene model, $a_1 = 1/2$, $a_2 = 1$. For white noise $\xi(t)$, the Fokker-Planck equation corresponding to (6.11) has the following form

$$\frac{\partial P(x,t)}{\partial t} = \left\{ -\frac{\partial}{\partial x} \left[\frac{1}{2} - x + a_3 x (1 - x) \right] + \right.$$

$$\left. D \frac{\partial}{\partial x} x (1 - x) \frac{\partial}{\partial x} x (1 - x) \right\} P(x,t) \tag{6.12}$$

with the stationary solution

$$P_{st}(x) = N (1 - x)^{-1 - a_3/D} x^{-1 + a_3/D} \exp\left(-\frac{1}{2Dx(1 - x)} \right). \tag{6.13}$$

For the symmetric case, $a_3 = 0$, $P_{st}(x)$ has only one maximum at $x = 1/2$ for $0 < x < 1$ if $D < 2$. But if $D > 2$, this point becomes a minimum and two maxima occur at

$$x = \frac{1}{2} \left(1 \pm \sqrt{1 - \frac{2}{D}} \right). \tag{6.14}$$

The value $D = 2$ defines the noise-induced phase transition analogous to equilibrium phase transitions [92]. For $a_3 \neq 0$, the qualitative picture is the same, but the transition occurs at $D = D(a_3)$.

Finally, for Ornstein-Uhlenbeck noise $\xi(t)$, quite complicate approximate calculations show [93] that the transition described above occurs at $D = 2(1 - \tau)$. However, the increase of the dimensionless correlation rate τ may lead to a unimodal-bimodal transition even for $D < 2$.

6.1.3 *Dichotomous noise*

For dichotomous multiplicative noise where $\xi(t)$ takes two values $\pm\sigma$ with a rate of jumps λ between these two states, one can solve Equation (6.3) for general forms of the deterministic part. For simplicity, consider the case of only multiplicative noise,

$$\frac{dx}{dt} = f(x) + g(x)\xi(t) \tag{6.15}$$

with $f(x)$ and $g(x)$ being arbitrary non-linear functions of x, and $\xi(t)$ is dichotomous noise. The Fokker-Planck equations for the distribution functions $P(x,t,\sigma)$ and $P(x,t,-\sigma)$ equivalent to the Langevin Equation (6.15) have the following form [94]

$$
\begin{aligned}
\frac{\partial P(x,t,\sigma)}{\partial t} &= -\frac{\partial}{\partial x}\left\{[f(x) + g(x)\sigma]P(x,t,\sigma)\right\} - \\
&\quad \frac{\lambda}{2}\left[P(x,t,\sigma) - P(x,t,-\sigma)\right], \\
\frac{\partial P(x,t,-\sigma)}{\partial t} &= -\frac{\partial}{\partial x}\left\{[f(x) - g(x)\sigma]P(x,t,-\sigma)\right\} + \\
&\quad \frac{\lambda}{2}\left[P(x,t,\sigma) - P(x,t,-\sigma)\right].
\end{aligned}
\tag{6.16}
$$

The stationary $(t \to \infty)$ solution for the probability density $P_{st}(x) = P(x,t,\sigma) + P(x,t,-\sigma)$ is easily obtained from (6.16)

$$P_{st}(x) = N\frac{|g(x)|}{D_{\text{eff}}(x)}\exp\left(\int^x \frac{f(y)}{D_{\text{eff}}(y)}dy\right) \tag{6.17}$$

where $D_{\text{eff}}(x)$ is an "effective diffusion coefficient" given by

$$D_{\text{eff}}(x) = \frac{\sigma^2}{2\lambda}\left[g(x) + \frac{f(x)}{\sigma}\right]\left[g(x) - \frac{f(x)}{\sigma}\right] \tag{6.18}$$

which has to be positive for the stable solution of Equation (6.15).

The extrema of $P_{st}(x)$ are the solutions of the following equation

$$f - \frac{\sigma^2}{2\lambda}g\frac{dg}{dx} + \frac{1}{\lambda}\left(2f\frac{df}{dx} - \frac{f^2}{g}\frac{dg}{dx}\right)_{x=x_e} = 0. \tag{6.19}$$

For Equation (6.3), $f(x) = ax - bx^2$ and $g(x) = x$. Therefore, the general form (6.17) of $P_{st}(x)$ reduces to [95]

$$P_{st}(x) = Nx^{[\lambda a/(\sigma^2-a^2)]-1}|bx+\sigma-a|^{-[\lambda/2(\sigma-a)]-1}|bx-\sigma-a|^{-[\lambda/2(\sigma+a)]-1}, \tag{6.20}$$

and the equation for the extrema is

$$bx - a + \frac{\sigma^2}{2\lambda} - \frac{2}{\lambda}(a - bx)(a - 2bx) + \frac{1}{\lambda}\left(ax - bx^2\right)^2 = 0. \tag{6.21}$$

When passing from dichotomous to white noise ($\sigma^2 \to \infty$, $\lambda \to \infty$ at $\sigma^2/\lambda = 2D$), equation (6.21) reduces to (6.9) with $\kappa = 0$, as it should.

6.1.4 *Poisson white noise*

After dichotomous noise, we consider the more realistic form of color noise, Poisson white noise, defined in (2.19)-(2.20). Both dichotomous and Poisson noise differ from white noise in that they both may fluctuate remaining restricted. This fact is of special importance for the decay rate b in (6.1), which has to be positive in order to get the stable solutions of Equation (6.1). We here bring the results of the appropriate calculations [96].

Fluctuations of the growth rate: $a(t) = a_0 + \xi(t)$.

For $\lambda w < a_0$, $a(t)$ remains positive and the stationary probability distribution

$$P_{st}(x) = N(x - x_1) x^{[(a_0/wbx_1)-1/w-1]} x^{-(a_0/wbx_1)-1}; \quad x_1 = \frac{a_0 - \lambda w}{b} \tag{6.22}$$

is defined in the domain (x_1, ∞).

For $\lambda w > a_0$, $a(t)$ can take negative values, and the probability density

$$P_{st}(x) = N_1 x^{(a_0/wbx_1)-1} (x + x_1)^{-a_0}; \quad x_1 = \frac{\lambda w - a_0}{b} \tag{6.23}$$

is defined in $(0, \infty)$.

Fluctuations of the decay rate: $b(t) = b_0 + \xi(t)$.

For $b_0 > \lambda w$, $b(t)$ is always positive and the stationary probability distribution is

$$P_{st}(x) = N \exp\left(-\frac{1}{wx}\right) x^{[(a-b_0x_1)/wax_1-1]} (x_1 - x)^{(b_0x_1-a)/wax_1};$$
$$x_1 = \frac{a}{b_0 - \lambda w}. \tag{6.24}$$

For $b_0 < \lambda w$, the situation is more complicated. Although the formal solution exists in the region $(0, \infty)$, it tends to infinity as $1/x^2$ at $x \to 0$, which has no physical meaning. Then, fluctuations in the parameter b with the Stratonovich interpretation of multiplicative noise lead to instability. The Ito interpretation produces a different result [97].

6.2 Response to a periodic force (noise-enhanced stability)

In the previous section, we considered the influence of multiplicative noise on Equation (6.1). Here we analyze the influence of a periodic force and

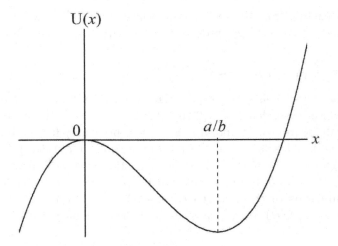

Fig. 6.1   A single-well potential with an unstable point at $x = 0$ and a metastable point at $x = a/b$. [Reprinted figure with permission from Ref. 98. Copyright (1992) by the American Physical Society.]

the additive noise which in the birth-death model represents a periodical and random influence of an environment [98].

Equation (6.1) with force $F \equiv -\frac{dU}{dx} = ax - bx^2$ describes the (over-damped) motion of a particle in the potential $U = -\frac{ax^2}{2} + \frac{bx^3}{3}$. This potential is characterized by two equilibrium points, one metastable, $x = a/b$, and one unstable, $x = 0$, as illustrated in Figure 6.1. In the presence of noise, a particle always escapes from the potential well, and we are interested in the effect of noise strength D on the mean escape time in the presence of an external periodic force. Let us start, however, from the deterministic equation obtained from (6.1) by adding a periodic force

$$\frac{dx}{dt} = ax - bx^2 + A\cos(\Omega t). \tag{6.25}$$

Although a full solution of this equation can be found in terms of Mathieu functions, we restrict ourselves to a qualitative analysis. An escape from the potential well depends on the initial position of a particle $x\,(t = 0)$, the amplitude A, and the frequency Ω of an external field. If A is smaller than the barrier height, $A < a^3/6b^2$, the particle will never leave the well provided that the initial position satisfies

$$x\,(t = 0) \geq \frac{a}{2b}\left(1 - \sqrt{1 + \frac{4Ab}{a^2}}\right). \tag{6.26}$$

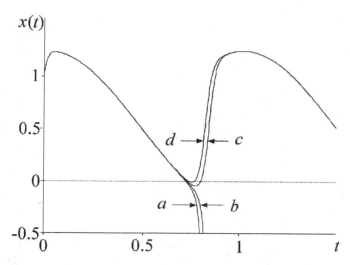

Fig. 6.2 Numerical solution of the dimensionless equation (6.25) for parameters $A = 0.3$, and frequencies a: $\Omega = 0.13196$, b: $\Omega = 0.13197$, c: $\Omega = 0.13198$, and d: $\Omega = 0.13199$. [Reprinted figure with permission from Ref. 98. Copyright (1992) by the American Physical Society.]

This follows from the fact that the upper bound of A ensures that for a fixed t, the right hand side of Equation (6.25) has two real roots, which is equivalent to the statement that the particle will remain trapped between the two values of x that correspond to these roots.

It is intuitively clear that the larger the amplitude of the field, the easier it is for the particle to escape from the well, but the extreme sensitivity of the escape to the frequency Ω of the field is much less obvious. Figure 6.2 shows curves $x(t)$ for several values of the frequency. These curves illustrate the fact that escape can occur at the end of any integer number of the cycle time $T = 2\pi/\Omega$ associated with the periodic term in (6.25). The results are extreme sensitive to changes in the frequency. Indeed, a change of 10^{-6} in the frequency determines whether the particle will escape after the first cycle. In Figure 6.3, we show the joint action of A and T on the escape from the well. For all values of A and T located above the upper curve, escape occurs even before the end of the first cycle. The uppermost curve corresponds to systems in which the particle escapes at the end of the first cycle. The curve just below corresponds to systems in which the particle escapes just at the end of the second cycle, etc. The points below the lowest curve in Figure 6.3 correspond to systems in which escape does not occur.

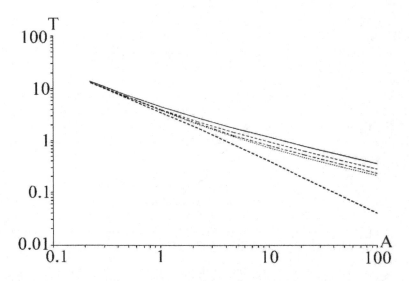

Fig. 6.3   Separatrics in the (A, T) plane, The region above the uppermost curve corresponds to the region in which a particle "escaprd" before the end of the first cycle. The region below the lowest curve shows the states in which a particle never escapes, and the intermidiate region corresponds to state in which a particle escapes after a finite number of cycles. [Reprinted figure with permission from Ref. 98. Copyright (1992) by the American Physical society.]

Let us now consider Equation (6.1) supplemented with a periodic force and additive noise

$$\frac{dx}{dt} = ax - bx^2 + A\cos\left(\Omega t\right) + \xi\left(t\right).\qquad(6.27)$$

The result of the numerical solution of Equation (6.27) for the escape time $\langle\tau\rangle$ is shown in Figure 6.4 for the initial condition $x\left(t=0\right)\ =\ 1$. These graphs look quite contra-intuitive. Indeed, at not too small A, both periodic force and noise favour escape, and it is believed that these two factors act together. However, as one can see from Figure 6.4, adding noise to a periodically driven system will increase (!) the escape time for some noise strength. This effect has been named [99] noise-enhanced stability NES)[1]. Noise-enhanced stability has now been found experimentally [99].

---

[1] In Hebrew "ness" means "miracle". As is well-known, to live in Palestine and perform miracles is quite dangerous... . Another dangerous prediction is that the addition of a periodic force (wars and the after-war children boom) to the birth-death equation, may, at some resonance conditions, increase the time it takes to reach population extinction $x = 0$; i.e., it may be useful for the survival of the species.

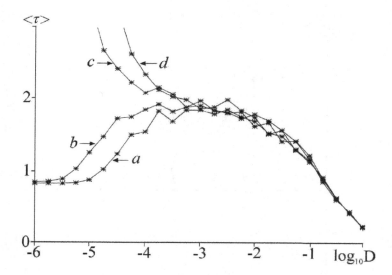

Fig. 6.4 The average escape time as a function of the logarithm of the noise strength for $A = 0.3$, and the same frequencies as used in Fig. 4. [Reprinted figure with permission from Ref. 98. Copyright (1992) by the American Physical Society.]

6.3 Piece-wise model of a metastable state

The effect of noise-enhanced stability (NES) can be explained quantitatively [100] by the example of a piece-wise linear potential $U(x)$ for $x > 0$ (Figure 6.5)

$$U(x) = \begin{cases} 0, & 0 < x < L \\ E - k(x - L), & L < x < b \end{cases} \qquad (6.28)$$

where E is the height of the potential barrier and b is the absorbing boundary. The states at $0 < x < L$ are metastable, and those at $L < x < b$ are unstable. The overdamped periodically driven motion of a particle in the potential (6.28) is described by the following equation

$$\frac{dx}{dt} = -\frac{dU}{dx} + A\sin(\Omega t) + \xi(t) \qquad (6.29)$$

with white noise $\xi(t)$ of strength D.

Consider first the time-independent potential ($A = 0$). If the initial position of a particle $x_0 = x(t = 0)$ is unstable, $L < x_0 < L + E/k$, the average escape time grows in the presence of noise since the particle may jump into the potential well. For very weak noise, the probability of such

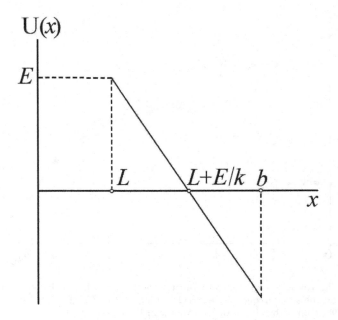

Fig. 6.5   The triangle piece-wise potential described by Equation (6.28).

jumps is very low. Just due to these jumps, the particle will be trapped in the well for a long time (noise enhanced stability).

Let us turn to the case $A \neq 0$. For $x_0 = 0$, the particle at $0 < x < L$ will move according to the equation $x(t) = (A/\Omega)[1 - \cos(\Omega t)]$. If $2A/\Omega < L$, the particle will always remain inside the region $(0, L)$. However, if $2A/\Omega > L$, the particle surmounts the region $(0, L)$, and its position will change with time as

$$x(t) = \frac{A}{\Omega}[1 - \cos(\Omega t)], \text{ for } 0 < t < t_1 \quad 0 < x(t) < L, \qquad (6.30)$$

and

$$x(t) = k(t - t_1) + \frac{A}{\Omega}[1 - \cos(\Omega t)], \text{ for } t > t_1 \quad L < x < b \qquad (6.31)$$

where t_1 is the time at which the particle crosses the point $x = L$,

$$t_1 = \frac{1}{\Omega}\arccos\left(1 - \frac{\Omega L}{A}\right). \qquad (6.32)$$

NES occurs if at time $t_2 = T/2 = \pi/\Omega$, when the periodic force changes it sign, tending to return the particle to the region $(0, L)$, and the particle is still located inside the interval $(0, b)$, $x(t_2) < b$. Using (6.32) and 6.31), the latter inequality can be rewritten as

$$\frac{2A}{b} + \frac{k}{b} \left[\pi - \arccos \left(1 - \frac{\Omega L}{A} \right) \right] < \Omega \qquad (6.33)$$

which, in addition to the previously used inequality

$$\Omega < \frac{2A}{L}, \qquad (6.34)$$

defines the conditions to the appearance of NES. Just these two inequalities define the regions in the $A - T$ plane explained previously and shown in Figure 6.3.

The onset of NES has been considered in [100] also in circumstances where a periodic force in (6.28) has been replaced by multiplicative dichotomous noise.

6.4 Rectangular potential barrier (stabilization of metastable state)

In the previous section, we considered the influence of noise on the escape time of a particle in a metastable state, and found that noise may enhanced the stability of a particle which otherwise will leave a metastable state under the action of a periodic signal. Here we consider slightly different situation involving a particle driven by noise from its metastable position, showing that an additional periodic signal (like noise in the previous chapter) is able to increase the stability of a metastable state[2].

As it well known [55], a simple pendulum is stable (metastable) in the vertically downward (upward) position. One can, however, stabilize a metastable position by applying high-frequency parametric oscillations of a pendulum (the "Kapitza pendulum"). As was shown both numerically [101] and analytically [102], the "dynamic stabilization" of a pendulum can be also achieved by adding an additive, and not multiplicative, periodic

---

[2]Rich become "richer" - this is the main slogan of the theory of self-organized criticality in growing networks, very popular nowadays, where a new site prefers to be connected with sites which already have many connections and become thereby even "richer". In this section we are aware of metastable states, which are "poor" - less stable and containing fewer particles than the stable states. Our aim is to find what are the possibilities for stabilization of metastable states, how to make them "richer". Our main slogan "poor become richer" is probably less realistic, but certainly more democratic.

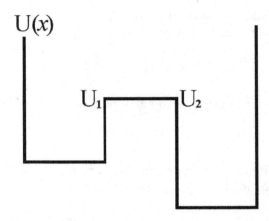

Fig. 6.6   The simplest form of the rectangular piece-wise bistable potential.

field. Incidentally, not only does the metastable state becomes stable, but also the stable states become unstable[3].

The question arises how general is this phenomenon, and whether one can stabilize other metastable states as well. In order to perform analytical calculations, we consider in this section the simplest model of a rectangular potential shown in Figure 6.6. As was shown in the last 10-20 years, many fundamental properties of a particle moving in a nonlinear potential are generic, and not too sensitive to the details of the potential. Therefore, it is worthwhile to consider the simplest potential that allows an analytical solution, in addition to numerical simulations for more realistic potentials. Our model (Figure 6.6) involves a particle subject to white noise of strength D moving in a piecewise double-well potential restricted by reflecting walls. The potential barriers U_1 and U_2 are different for the right (stable) and the left (metastable) states.

For the potential $U(x)$ shown in Figure 6.6, $dU/dx = 0$ everywhere except at the matching points. The Fokker-Planck equation, which can be written in the form

$$\frac{\partial P}{\partial t} = \frac{\partial}{\partial x}\left(\frac{dU}{dx}P + D\frac{\partial P}{\partial x}\right) \equiv -\frac{\partial J}{\partial x}, \qquad (6.35)$$

reduces to a simple diffusion equation, namely, one has to solve Equation (6.35) in each of three different regions of $U = $ const., and then ensure

---

[3]The previously declared slogan "poor become richer" is now complemented by the even more revolutionary slogan "rich become poorer".

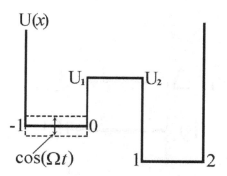

Fig. 6.7 Square double-well potential with an oscillating left well. [Reprinted figure with permission from Ref. 104. Copyright (1999) by the American Physical Society.]

the continuity of P and J on the boundaries of these regions. Continuity of probability current J, which according to Equation (6.35) can be written as $J = -D \exp(-U/D) \frac{d}{dx} [\exp(U/D) P]$, means that at points x_i of the jumps of potentials, one gets

$$\exp\left(\frac{U(x_i+0)}{D}\right) P(x_i+0) = \exp\left(\frac{U(x_i-0)}{D}\right) P(x_i-0), \qquad (6.36)$$

$$\frac{\partial P(x_i+0)}{\partial x} = \frac{\partial P(x_i-0)}{\partial x}.$$

The matching conditions (6.36) have to be complemented by reflected boundary conditions at the positions $\pm L$ of the walls, $\partial P(\pm L, t)/\partial x = 0$. It is easy to check [103] that the normalized stationary probability distribution function $P_{st}(x)$ in the absence of an external field, has the following form in each of three regions shown in Figure 6.6,

$$P_{1,st} = \frac{\exp(U_1/D)}{2a + (L-a)[\exp(U_1/D) + \exp(U_2/D)]},$$

$$P_{2,st} = \frac{1}{2a + (L-a)[\exp(U_1/D) + \exp(U_2/D)]}, \qquad (6.37)$$

$$P_{3,st} = \frac{\exp(U_2/D)}{2a + (L-a)[\exp(U_1/D) + \exp(U_2/D)]}.$$

Since the two wells have equal width, the equilibrium population of the left (metastable) state is smaller than that of the right (stable) state by the ratio $\exp[(U_2 - U_1)/D]$.

An external force can be chosen as a periodic force acting on the well as is shown in Figure 6.7, or as a random force acting on the barrier (Figure

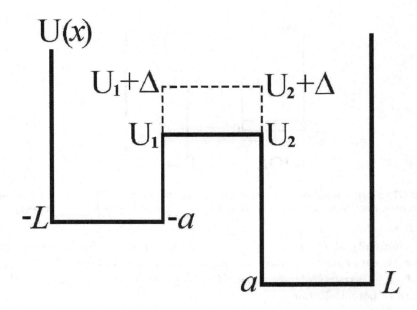

Fig. 6.8 Square double-well potential subject to the dichotomous fluctuations of the barrier height. [Reprinted figure with permission from Ref. 105. Copyright (2002) by the American Physical Society.]

6.8). It turns out that for both cases, an external force attempts to equalize the populations at $t \to \infty$ (stabilizing the metastable state), even reversing the populations of these states. Our choice of the periodic signal does not introduce an additional force in Equation (6.35). However, the periodic signal enters the matching conditions (6.36).

Starting from the periodic changes of the well (Figure 6.7) [104], we assume that the amplitude of the external field A is small, $A < D$. Accordingly, we seek the solution of Equation (6.35) in each region, $i = 1, 2, 3$, as

$$P_i = P_{i,st} + \sum_{l=1}^{\infty} \left(\frac{A}{D} \right)^l f_i^l (x, t) \tag{6.38}$$

where f_i^l is a periodic function of t which can be written in the following form

$$f_i^l = f_{i,0}^l + \sum_{k=1}^{\infty} \left\{ \left[f_{i,k}^l \exp\left(r_k x \right) + f_{i,k}^{-l} \exp\left(-r_k x \right) \right] + c.c. \right\} \exp(i\Omega k t), \tag{6.39}$$

$$r_k = \sqrt{\frac{i\Omega k}{D}}.$$

Due to the boundary conditions on the walls the first order in time-independent corrections vanish, therefore, we keep also the non-oscillating second-order terms $\left(\frac{A}{D}\right)^2 f_i$ in (6.38). Then, one gets twelve parameters $(P_{i,st}, f_i, \tilde{f}_i \, g_i; \, i = 1, 2, 3)$ that satisfy twelve equations (two on the wells and two on each of two matching points for each order in the small parameter A/D). Omitting all calculations, we present here only the time-independent populations (up to second order in $(A/D)^2$) of the right well, $n_{r,\infty}$, and the left (oscillating) well, $n_{l,\infty}$,

$$n_{l,\infty} = \int_{-1}^{0} P_1(x,t)\, dx = P_{1,st} + \left(\frac{A}{D}\right)^2 g_1 + \ldots \tag{6.40}$$

$$n_{r,\infty} = \int_{1}^{2} P_3(x,t)\, dx = P_{3,st} + \left(\frac{A}{D}\right)^2 g_3 + \ldots$$

where $P_{1,st}$ and $P_{3,st}$ are defined in (6.37), and

$$g_1 = -\left[1 + \exp(U_2/D)\right]^{-1}; \quad g_3 = -\left(P_{1,st} + P_{3,st}\right)\left[P_{1,st}/4 + 2\,\mathrm{Re}\,f_1\right];$$

$$\mathrm{Re}\,f_1 = \frac{1 + \exp(-U_2/D)}{2\exp\left[(U_1 + U_2)/D\right]} \frac{4H^2 + \sin^2(\alpha_2)\sinh^2(\alpha_2)}{4H^2 - \left[\sinh^2(\alpha_2) - \sin^2(\alpha_2)\right]};$$

$$H = \frac{\left[1 + \exp(U_1/D) + \exp(U_2/D)\right]\left[\sinh^2(\alpha_1) + \cos^2(\alpha_1)\right]}{\exp(U_1/D) + \exp(U_2/D)} +$$

$$\frac{\sinh^2(\alpha_2) - \sin^2(\alpha_2)}{4};$$

$$\alpha_n = \sqrt{\frac{n^2\Omega}{2D}}. \tag{6.41}$$

With the help of an external periodic field, one can increase the population of the left (metastable) state, or even reverse the populations. The latter, $n_{l,\infty} > n_{r,\infty}$, will occur when

$$P_{3,st} - P_{1,st} < \left(\frac{A}{D}\right)^2 (g_1 - g_3) \tag{6.42}$$

which gives

$$\left(\frac{A}{D}\right)^2 \frac{1 + 2\exp(U_2/D)}{\exp(U_2/D) - \exp(U_1/D)}\left[-\frac{P_{1,st}}{4} - 2\,\mathrm{Re}\,f_1\right] > 1. \tag{6.43}$$

The last inequality is obeyed when the factor in front of the brackets is large, and the expression in brackets is positive. The former occurs when $A > \sqrt{U_2 - U_1}$, and the latter holds for not too small frequencies.

Another form of stabilization of a metastable state is shown in Figure 6.8, where the potential barrier undergoes dichotomous fluctuations of height Δ with flipping rate λ, i.e., the barrier for the right (left) well changes randomly between heights U_1, $U_1 + \Delta$ (U_2, $U_2 + \Delta$) [105]. The question is whether the dichotomous fluctuations of the potential barrier are able (like an external periodic field) to increase the stationary $(\partial P/\partial t = 0)$ population of a metastable state. Two probability functions $P^\pm (x)$, which define the probability to be at position x when the potential is $U_1 + \Delta$ or U_1, are related by the following equations

$$D\frac{d^2 P^\pm}{dx^2} + \frac{\lambda}{2}\left(P^\mp - P^\pm\right) = 0. \tag{6.44}$$

The solutions of equations (6.44) have the following form

$$P^\pm_{1,2,3} = a_{1,2,3} + b_{1,2,3}\ x \pm c_{1,2,3}\sinh\left(\sqrt{\frac{\lambda}{D}}x\right) \pm d_{1,2,3}\cosh\left(\sqrt{\frac{\lambda}{D}}x\right) \tag{6.45}$$

where the subscripts $1, 2, 3$ relate to the three regions shown in Figure 6.7. The coefficients in Equation (6.45) have to be found from the matching conditions (6.36) and boundary conditions. Once these coefficients are found, one can calculate the populations $n_{1,\infty}$ of the metastable and $n_{3,\infty}$ of the stable state,

$$n_{1,\infty} = \int_{-L}^{-a}\left[P_1^+(x) + P_1^-(x)\right]dx; \quad n_{3,\infty} = \int_a^L\left[P_3^+(x) + P_3^-(x)\right]dx. \tag{6.46}$$

Instead of writing the cumbersome formulas for $n_{1,\infty}/n_{3,\infty}$ [105], we shall bring their limiting forms for small and large amplitudes of noise, $\Delta/D \lessgtr 1$, namely, for $\Delta/D < 1$,

$$\frac{n_{1,\infty}}{n_{3,\infty}} = \exp\left[(U_2 - U_1)/D\right] \times \tag{6.47}$$

$$\left\{1 + \frac{2\left[\exp\left(-\frac{U_2}{D}\right) - \exp\left(-\frac{U_1}{D}\right)\right]\coth\left[\sqrt{\frac{\lambda}{D}}(L - a)\right]}{1 + 16B_1 B_2 + 4\left(B_1 + B_2\right)\coth\left(2\sqrt{\frac{\lambda}{D}}a\right)}\left(\frac{\Delta}{D}\right)^2 + ...\right\}$$

and for $\Delta/D > 1$,

$$\frac{n_{1,\infty}}{n_{3,\infty}} = \exp\left[(U_2 - U_1)/D\right] \times \tag{6.48}$$

$$\left\{1 + \frac{2\left[\exp\left(-\frac{U_2}{D}\right) - \exp\left(-\frac{U_1}{D}\right)\right]\coth\left[\sqrt{\frac{\lambda}{D}}(L - a)\right]\tanh\left(\sqrt{\frac{\alpha}{D}}a\right)}{1 + 2B_1\tanh\left(2\sqrt{\frac{\lambda}{D}}a\right)} + ...\right\}$$

where

$$B_1 = \exp\left(-\frac{U_1}{D}\right) \coth\left[\sqrt{\frac{\lambda}{D}}\,(L-a)\right];$$ (6.49)

$$B_2 = \exp\left(-\frac{U_2}{D}\right) \coth\left[\sqrt{\frac{\lambda}{D}}\,(L-a)\right].$$

Equation (6.48) shows that the ratio of populations in the left (metastable) and right (stable) states increases with increasing strength of the noise or the noise rate. For some values of parameters the latter dependence is non-monotonic [105].

In conclusion, the population of a metastable state can be increased by the use of an external periodic field or by fluctuations of the barrier height.

Chapter 7

Overdamped Double-well Oscillator

Fluctuations of a double-well system driven by multiplicative $\xi(t)$ and additive $\eta(t)$ noise are described by the following equation

$$\frac{dx}{dt} = ax - bx^3 + x\xi(t) + \eta(t). \tag{7.1}$$

Equation (7.1) describes the motion of a particle in the potential $U(x) = -ax^2/2 + bx^4/4$ with two potential minima located at $x = \pm\sqrt{a/b}$ and a maximum at $x = 0$ (Figure 7.1). The solution of the deterministic Equation (7.1), with $\xi = \eta = 0$, is

$$x(t) = \frac{x(t=0)}{\sqrt{\frac{b}{a}x^2(t=0) + \left[1 - \frac{b}{a}x^2(t=0)\right]\exp(-2at)}} \tag{7.2}$$

which means that at $t \to \infty$ a particle will stay in one of the minima $x = \pm\sqrt{a/b}$, depending on the sign of $x(t=0)$. Our aim is to study Equation (7.1) with noise included. Among many application described by this Equation are: statistical properties of the dye-laser laser light [106],[107], hydrodynamic instabilities [108], autocatalytic chemical reactions [109], current instabilities in parametrically excited circuits [110], and electrodynamic instabilities in nematic liquid crystals [92].

7.1 Steady state

7.1.1 *White noise*

If the noise $\xi(t)$ and $\eta(t)$ in (7.1) are correlated white noise described by (6.4), the analysis of the stationary solutions of Equation (7.1) is quite similar [111] to that presented in section 6.2 for the birth-death equation.

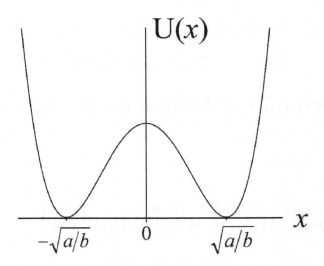

Fig. 7.1   A double-well potential with two minima located at $\pm\sqrt{a/b}$ and a maximum at $x = 0$.

The stationary distribution function $P_{st}(x)$ for the Langevin Equation (7.1) has the following form for $0 \leq \kappa < 1$ [111]

$$P_{st}(x) = N\left(Dx^2 + 2\kappa\sqrt{D\alpha}x + \alpha\right)^{C-1/2} \times$$

$$\exp\left\{f(x) + \frac{E}{\left[(1-\kappa^2)D\alpha\right]^{1/2}}\arctan\frac{Dx + \kappa\sqrt{D\alpha}}{\left[(1-\kappa^2)D\alpha\right]^{1/2}}\right\} \quad (7.3)$$

where

$$f(x) = 2\kappa\sqrt{\frac{\alpha}{D}}\frac{bx}{D} - \frac{bx^2}{2D}, \quad E = \kappa\sqrt{\frac{\alpha}{D}}\left[(4\kappa^2 - 1)\frac{b\alpha}{D} - a\right], \quad (7.4)$$

$$C = \frac{b\alpha + aD - 4b\alpha\kappa^2}{2D^2}.$$

For perfect correlated noise, $\kappa = 1$,

$$P_{st}(x) = N\left(\sqrt{\alpha} + \sqrt{D}x\right)^{2C(\kappa=1)-1}\exp f(\kappa=1) + \frac{E(\kappa=1)}{Dx + \sqrt{D\alpha}}. \quad (7.5)$$

The extrema of $P_{st}(x)$ obey the following equation

$$bx^3 + (D - a)x + \kappa\sqrt{D\alpha} = 0; \quad \text{for } 0 \leq \kappa < 1. \quad (7.6)$$

For non-correlated noise, $\kappa = 0$, Equation (7.6) has two roots, $x = 0$ and $x = \sqrt{(a-D)/b}$ [112]. For small intensity of noise, $D < a$, $P_{st}(x)$ has a single maximum at $x = \sqrt{(a - D)/b}$ which is shifted to smaller values of x when D increases. For $D > a$, the maximum of $P_{st}(x)$ disappears and P_{st} goes to infinity at $x \to 0$.

Some interesting results follow from the analysis of Equations (7.3)–(7.6):

1. The presence of correlation between noise, $\kappa \neq 0$, leads to essential changes in the steady state distribution, namely, the distribution function $P_{st}(x)$ which had a symmetric bimodal structure at $\kappa = 0$, becomes non-symmetric for $\kappa \neq 0$.

2. As is well known [92], the extrema of $P_{st}(x)$ are shifted by multiplicative noise but not by additive noise. However, as one can see from (7.6), in the case of the correlation between these two sources of noise, the extrema of $P_{st}(x)$ are affected not only by multiplicative noise but also by additive noise.

3. Without noise, the steady state distribution has two minima and one maximum ($D = \alpha = 0$ in (7.6) gives $x = 0, \pm\sqrt{a/b}$), while the presence of noise may change this distribution from bimodal to unimodal. The latter is influenced by both multiplicative and additive noise and takes place when the three solutions of Equation (7.6) reduce to one solution, i.e., when

$$\frac{\kappa^2 D\alpha}{4} + \frac{(D - a)^3}{27b} = 0. \tag{7.7}$$

4. For the case of perfectly correlator noise, $\kappa = 1$, the steady state distribution (7.5) exhibits a divergence at $x = -\sqrt{\alpha/D}$, and the function $P_{st}(x)$ exhibits a different shape of divergence for $b\alpha/aD > 1$ and $b\alpha/aD < 1$.

5. The first moment $\langle x \rangle_{st}$ decreases with increasing κ. The effect of κ on $\langle x \rangle_{st}$ is more pronounced for $D \geq \alpha$ than for the case of $D < \alpha$.

A slightly more complicated case, where the delta-correlation between noise (6.4) is replaced by the exponential form correlation with the correlation time as an additional parameter, has been considered in [113], [114].

7.1.2   Dichotomous noise

For the Langevin equation with multiplicative noise

$$\frac{dx}{dt} = ax - bx^3 + \xi(t)x \tag{7.8}$$

one can get an explicit formal solution by the change of variables indicated above in (3.25)-(3.26), namely, in terms of the new variable $y = x^{-2}$,

Equation (7.8) takes the following form

$$\frac{dy}{dt} = 2b - 2\left[a + \xi\left(t\right)\right] y. \tag{7.9}$$

This linear equation can be solved exactly which allows one to find the dynamic (time-dependent) moments $\langle y^n\left(t\right)\rangle$, i.e. the inverse moments of $x\left(t\right)$. Although it does not help to solve the original Equation (7.8), this approach serves [115] as a useful test for the validity of an approximate solution of (7.8) for small values of the correlation time.

Let us restrict ourselves to multiplicative dichotomous noise in (7.8). The analysis similar to (6.16)-(6.21) can be performed for the Langevin Equation (7.8), which gives the stationary distribution function

$$P_{st}(x) = Nx^{-1+\lambda a/(\sigma^2-a^2)}|a - \sigma - bx^2|^{-1+\lambda/2(a-\sigma)}|a + \sigma - bx^2|^{-1+\lambda/2(a+\sigma)} \tag{7.10}$$

and the extrema of (7.10)

$$a - bx^2 - \frac{\sigma^2}{\lambda} + \frac{2}{\lambda}\left(a - bx^2\right)\left(a - 3bx^2\right) - \frac{1}{\lambda}\left(ax - bx^3\right)^2 = 0. \tag{7.11}$$

Linear stability analysis and the calculation of equilibrium moments have been performed for this case in [116].

7.2   Eigenfunction expansion of the Fokker-Planck equation

In addition to the stationary solution of the Fokker-Planck equation, one can find the full solution of this equation by introducing the following assumption,

$$P\left(x,t\right) = P\left(x\right)\exp\left(-\varepsilon t\right) \tag{7.12}$$

which changes the type of the Fokker-Planck differential equation from a partial to an ordinary differential equation [109].

Taking into account (7.12), the time-dependent Fokker-Planck equation, which corresponds to the Langevin Equation (7.8), has the following form

$$\frac{d^2}{dx^2}\left[x^2 P\left(x\right)\right] - \frac{d}{dx}\left\{\left[\left(1 + \frac{a}{D}\right) x - \frac{bx^3}{D}\right] P\left(x\right)\right\} = -\frac{\varepsilon}{D} P\left(x\right). \tag{7.13}$$

Solutions of this Equation have a different form for $\varepsilon \le a^2/4D$ and $\varepsilon > a^2/4D$. In the former case, the eigenfunctions of Equation (7.13) are

$$P_n\left(x\right) = x^{-1+(a/2D)-2n}\exp\left(-\frac{bx^2}{2D}\right){}_1F_1\left(-n, 2v_n, \frac{bx^2}{2D}\right) \tag{7.14}$$

where $_1F_1$ is the confluent hypergeometric function [117],

$$2v_n = 1 + \frac{a}{2D} - 2n > 0 \tag{7.15}$$

and the eigenvalues spectrum

$$\varepsilon_n = 4nD\left(\frac{a}{2D} - n\right). \tag{7.16}$$

The distribution function (7.14) has a singularity at $x = 0$ for $-1 + a/D. -2n < 0$ which disappears if one includes additive noise in the original Langevin equation.

For $\varepsilon > a^2/4D$, the eigenfunctions are

$$P_n(x) = x^{-1+a/2D} \exp\left(-\frac{bx^2}{2D}\right) [f(\mu) + f(-\mu)] \tag{7.17}$$

with

$$f(\mu) = \frac{\Gamma(-2\mu)}{\Gamma\left(\frac{1}{2} - \mu - \nu\right)} x^{2\mu} \, _1F_1\left(\mu - \nu + \frac{1}{2}, 2\mu + 1, \frac{bx^2}{2D}\right), \tag{7.18}$$

$$\mu = \frac{i}{\sqrt{4D}}\sqrt{\varepsilon - \frac{a^2}{4D}}; \ \nu = \frac{1}{2}\left(\frac{a}{2D} + 1\right)$$

and eigenvalues cover the continuous part of the spectrum for all real values of ε satisfying the condition $\varepsilon > a^2/4D$.

Note the remarkable property of the eigenvalue spectrum which consists of both discrete and continuum branches.

7.3 Matrix continued fraction method

In previous section, we considered the general method of solution the Fokker-Planck equation. However, in many cases a system is subjected to a periodic force, and it is then worthwhile to use matrix continued fractions. The typical equation of motion with white noise $\xi(t)$ has the following form

$$\frac{dx}{dt} = f(x) + A\sin(\Omega t) + \xi(t) \tag{7.19}$$

where $f(x)$ is the arbitrary function of x. The Fokker-Planck equation corresponding to the Langevin Equation (7.19) is

$$\frac{\partial P(x,t)}{\partial t} = -\frac{\partial}{\partial x}\left\{[f(x) + A\sin(\Omega t)]P(x,t)\right\} + D\frac{\partial^2 P(x,t)}{\partial x^2}. \tag{7.20}$$

As was shown in chapter 5.2, for the linear function $f(x) = ax$, these equations possess exact solutions for the first two moments. In the general case, one can use the periodicity in time of the coefficients in the differential Equation (7.20), which means [118] that (7.20) possesses Floquet-type solutions

$$P(x,t) = \exp(\alpha t) p_\alpha(x,t) \qquad (7.21)$$

where the functions $p_\alpha(x,t)$ are periodic in time, i.e., $p_\alpha(x,t+T) = p_\alpha(x,t)$, where $T = 2\pi/\Omega$. Expanding the periodic function $p_\alpha(x,t)$ in a Fourier series

$$p_\alpha(x,t) = \sum_{n=-\infty}^{n=\infty} c_n^\alpha(x) \exp(i\Omega t), \qquad (7.22)$$

and substituting (7.21) and (7.22) into (7.20), one gets

$$\left[-\frac{df}{dx} - f\frac{d}{dx} + D\frac{d^2}{dx^2} - in\Omega - \alpha \right] c_n^\alpha(x) - \frac{iA}{2} \left[c_{n+1}^\alpha(x) - c_{n-1}^\alpha(x) \right] = 0. \qquad (7.23)$$

It is convenient [119] to expand the coefficients $c_n^\alpha(x)$ into the complete orthogonal set of eigenfunctions

$$c_n^\alpha(x) = \sum_{m=0}^\infty c_n^m \psi_m(x). \qquad (7.24)$$

As an example, for the double-well potential $f(x) = x - x^3$, it is convenient to use the Hermit polynomials $H_n(\rho x)$,

$$\psi_m(x) = \sqrt{\frac{\rho}{n!2^n\sqrt{\pi}}} \exp\left(-\frac{\rho^2 x^2}{2} \right) H_n(\rho x) \qquad (7.25)$$

where ρ is an arbitrary parameter. Inserting (7.24) and (7.25) into (7.23) and using the orthogonality and recursion properties of the Hermit polynomials, one gets

$$\sum_{\overline{m}=0}^\infty \left[Q_{m,\overline{m}}(n) c_n^{\overline{m}} + Q_{m,\overline{m}}^+ c_{n+1}^m + Q_{m,\overline{m}}^- c_{n-1}^m \right] = 0 \qquad (7.26)$$

with the complex matrices

$$Q_{m,\overline{m}}(n) = \left(-\frac{df}{dx} - f\frac{d}{dx} + D\frac{d^2}{dx^2} \right)_{m,\overline{m}} - in\Omega\delta_{m,\overline{m}} - \alpha\delta_{m,\overline{m}}, \qquad (7.27)$$

$$Q_{m,\overline{m}}^+(n) = -Q_{m,\overline{m}}^-(n) = -\frac{iA}{2}B_{m,\overline{m}}.$$

The cumbersome expressions for the matrices $\left(-\frac{df}{dx} - f\frac{d}{dx} + D\frac{d^2}{dx^2} \right)_{m,\overline{m}}$ and $B_{m,\overline{m}}$ are given in [119].

Hence the problem is reduced to analysis of the tridiagonal recursion relations

$$Q_n(\alpha)\,\overline{c_n} + Q_n^+\overline{c_{n+1}} + Q_n^-\overline{c_{n-1}} = 0 \qquad (7.28)$$

which appear in different problems [120]. The solution of (7.28) can be found by introducing the rates

$$S_n = \frac{\overline{c_{n+1}}}{\overline{c_n}}. \qquad (7.29)$$

Inserting (7.29) into (7.28) leads to

$$\overline{c_n} = -\frac{Q_n^-}{Q_n + Q_n^+ S_n}\,\overline{c_{n-1}} \equiv S_{n-1}\overline{c_{n-1}}. \qquad (7.30)$$

Comparing (7.29) and (7.30) yields a recurrent relation for S_n,

$$S_n = -\frac{Q_{n+1}^-}{Q_{n+1} + Q_{n+1}^+ S_{n+1}} \qquad (7.31)$$

which, starting from $S_0 = c_1/c_0$, can be used again and again. Finally, one gets

$$c_n = S_{n-1}S_{n-2}...S_0c_0. \qquad (7.32)$$

Another method for solving Equation (7.28) is the transition from a difference to a differential equation by replacing the integers n by the continuous variable and using a Taylor expansion

$$\overline{c_{n+1}} - \overline{c_{n-1}} \cong 2\frac{d\overline{c_n}}{dn} + O\left(\frac{2}{3}\frac{d^3\overline{c_n}}{dn^3}\right). \qquad (7.33)$$

For some special form of the matrices $Q(\alpha)$ and Q^+, this method gives quite accurate results [121].

7.4 Mean first-passage time

A particle located in, say, the left minimal position, $x_1 = -\sqrt{a/b}$, of the potential shown in Figure 7.1 will stay there forever unles some external force, such as an external noise, causes the particle to leave this minimum and to jump to the right minimal position, $x_2 = \sqrt{a/b}$. For analysis of this and similar processes, let us define the mean first-passage time T in terms of probability that the random variable x starting from the value x_0 reaches the value x_1 for the first time in a time interval between t and $t + dt$. "For the first time" is taken to mean that after reaching the value x_1, the process is finished, a particle is no longer counted, i.e. the point x_1 is the absorbing point with vanishing probability $P(x_1, T) = 0$. The latter condition plays a role of the boundary condition for the Fokker-Planck equation.

It is not surprising, therefore, that the mean first-passage time T can be expressed [34] in terms of the solution of the Fokker-Planck equation corresponding to (7.1).

The probability that the first-passage time T from $x = x_1$ to $x = x_2$ is greater than t is defined by the so-called survival probability $S(t)$,

$$S(t) = \int_{x_1}^{x_2} P(x, t)\, dx. \tag{7.34}$$

All the statistics of the first-passage time can be computed from $S(t)$. For example, the mean free-passage time is defined as

$$T = \int_0^{\infty} S(t)\, dt. \tag{7.35}$$

In the case of two correlated sources of white noise (6.4), T is given by [122]

$$T = \int_{x_1}^{x_2} dx \left[\left(Dx^2 + 2\kappa\sqrt{D\alpha}\, x + \alpha \right) P_{st}(x) \right]^{-1} \int_{-\infty}^{x} dy\, P_{st}(y). \tag{7.36}$$

For noise with strengths D and α smaller than the potential barrier, the mean free-passage time becomes independent of the initial condition $x(t = 0)$ and of the final point x_2 (it is enough to reach the point $x = 0$ of potential maximum in order to reach the point x_2). The explicit expression for T [123] for $0 \leq \kappa < 1$, has the following form

$$T = \sqrt{2\pi} \left| \frac{aD}{b\alpha} - 2\kappa\sqrt{\frac{aD}{b\alpha}} + 1 \right|^{-C} \exp{-\frac{a}{2D}} + \frac{2b\kappa\sqrt{D\alpha}}{D^2} +$$

$$\frac{E}{D}\left(\arctan\frac{\kappa}{\sqrt{1 - \kappa^2}} - \arctan\frac{\kappa - \sqrt{aD/b\alpha}}{\sqrt{1 - \kappa^2}} \right) \tag{7.37}$$

and for $\kappa = 1$

$$T = \sqrt{2\pi} \left| 1 - \sqrt{\frac{aD}{b\alpha}} \right|^{-2C_{\kappa=1}} \exp\left(-\frac{a}{2D} + \frac{2b\sqrt{D\alpha}}{D^2} + \frac{E_{\kappa=1}}{D\sqrt{\alpha} - \alpha\sqrt{D}} \right) \tag{7.38}$$

where C and E are defined in (7.4).

Analysis of formulas (7.37)-(7.38) shows that [123]

1. For fixed α (D), T increases with increasing κ, but decreases when D (α) increases. The effect of κ on T is in good agreement with the transition

from a bimodal to unimodal structure of $P_{st}(x)$ as κ increases, so that the effect of κ is to make T decrease.

2. The mean first-passage time is biggest for $D = \alpha$, as compared with $D \gtrless \alpha$.

3. For the perfectly correlated noise $\kappa = 1$, the dependence of T on D and on α is quite different. When α is fixed, T increases with increasing D for $\alpha > D$, and decreases for $\alpha < D$. On the other hand, when D is fixed, T increases with increasing α for $\alpha < D$, and decreases for $\alpha > D$.

If the correlation time τ of the correlation between additive and multiplicative noise is nonzero, one can study the dependence of the mean first-passage time T on τ [124], [125].

7.5 Response to a periodic force (stochastic resonance)

In chapter 5 we considered stochastic resonance in linear systems. However, this phenomenon usually appears under the heading of "nonlinear phenomena", describing a nonlinear (usually bistable) system driven by a combination of an additive random and periodic forces

$$\frac{dx}{dt} = ax - bx^3 + \eta(t) + A\cos(\Omega t). \qquad (7.39)$$

The amplitude of the periodic force A in (7.39) is assumed to be sufficiently small that in the absence of noise this force is unable to move a particle from one well to the other. This condition is satisfied when A is smaller than the barrier height $\Delta U = U(0) - U\left(\sqrt{\frac{a}{b}}\right) = \frac{a^2}{4b}$, i.e., $A < \frac{a^2}{4b}$. Although this signal is unable by itself to transfer a particle through the barrier, it makes it easier for the random signal to induce a transfer. The latter occurs when the deterministic frequency Ω approaches a characteristic frequency of the system without the periodic force. The appropriate frequency to characterize the Equation (7.39) in the absence of a periodic force is the Kramers rate $r = r_0 \exp(-\Delta U/D)$, where r_0 has the dimensions of frequency. Most curious is the role of noise in this process. Too small noise cannot help to overcome the threshold (barrier height), while too strong noise destroys the signal. Therefore, an intermediate strength of the noise will be the most useful.

The manifestation of stochastic resonance can roughly be divided into two classes. The first relates to $x(t)$ at the characteristic frequency and can be expressed in terms of different functions that are calculated in terms of $x(t)$. The mathematical form of this relation is usually represented in terms of Fourier components of these quantities which behave non-monotonically when evaluated as a function of the noise strength D. A second group of physically interesting parameters which can show resonant behavior are characteristic times, such as the reciprocal of the switching time between the wells in the model defined by Equation (7.39) [126]. This can also depend non-monotonically of the strength D of the noise point Still another measure of stochastic resonance is the input energy, i.e., the work done by an external field, which turns out to be non-monotonic function of both noise strength [127] and frequency of an external field [128].

In principle, one can use stochastic resonance for the resolution of the long-standing engineering problem of the detection and amplification of weak signals embedded within a large noise background. Thereby, the weak signal increases at the expense of noise. Since noise is found everywhere in nature, the question arises if "weak" signals are actually weak[1].

There are several comprehensive reviews on stochastic resonance [129]– [132], containing the full analysis of Equation (7.39). Therefore, we have restricted ourselves to the case of two sources of noise,

$$\frac{dx}{dt} = ax - bx^3 + \xi(t)x + \eta(t) + a\cos(\Omega t). \qquad (7.40)$$

Let us start with the case of non-correlated white noise of strength D and α with $\kappa = 0$ in (6.4) [133]. We assume, as it usually done in the theory of stochastic resonance, that both amplitude A and frequency Ω are small enough. The smallness of amplitude means that in the absence of any noise, the periodic force is unable to force a particle to move from one well to the another, while the smallness of the frequency (the so-called adiabatic limit) allows the system to reach local equilibrium in the period of $2\pi/\Omega$. In the adiabatic limit, one can simply add the slowly changing periodic term as a constant into an expression obtained for the static potential, and get the quasi-steady-state distribution function $P_{st}(x,t)$,

$$P_{st}(x,t) = N \left|Dx^2 + \alpha\right|^{a/2D + b\alpha/2D^2 - 1/2} \times$$
$$\exp\left[-\frac{bx^2}{2D} + \frac{A}{\sqrt{D\alpha}}\arctan\left(\sqrt{\frac{D}{\alpha}}x\right)\cos(\Omega t)\right] \qquad (7.41)$$

The distribution function $P_{st}(x,t)$, according to (7.36), allows one to find the first-passage time T in terms of the first-passage time T_0 in the

---

[1]Pursuing these questions, one can think of far-reaching applications, such as the real danger of the "weak" radiation from our computers and cellular telephones, the ability of telepathic transition of "weak" signals, etc.

absence of a periodic signal

$$T = T_0 \exp \left[A \sqrt{\frac{D}{\alpha}} \arctan \left(\sqrt{\frac{aD}{b\alpha}} \right) \cos \left(\Omega t \right) \right] \qquad (7.42)$$

where

$$T_0 = \frac{\sqrt{2}\pi}{a} \left| 1 + \frac{aD}{b\alpha} \right|^{a/2D + b\alpha/2D^2} \exp \left(-\frac{a}{2D} \right). \qquad (7.43)$$

Equations (7.41) and (7.43) reduce to (7.3) and (7.37) with non-correlated noises, $\kappa = 0$.

Stochastic resonance is characterized by the so-called signal-to-noise ratio (SNR) which is defined in (5.31). For Equation (7.40), one gets [133]

$$SNR = \frac{\pi A^2}{4T_0 D\alpha} \left(\arctan \sqrt{\frac{aD}{b\alpha}} \right)^2 \left[1 - \frac{A^2 \left(\arctan \sqrt{\frac{aD}{b\alpha}} \right)^2}{2D\alpha \left(\Omega^2 T_0^2 + 1 \right)} \right]^{-1}. \qquad (7.44)$$

Analysis of (7.44) shows that effects of the multiplicative and additive noise strengths on the signal-to-noise ratio are opposed to each other, namely, the peak of SNR increases (decreases) with increase of D (α) when $\alpha(D)$ is maintained fixed. In both cases, the peak increases with increasing amplitude A of the periodic signal. However, for the fixed α (and not for the fixed D), the second maximum in SNR can appear when the amplitude of the periodic signal A is increased, replacing thereby a "single stochastic resonance" (SSR) by a "double stochastic resonance" (DSR). This phenomenon was found for the first time by the analysis of traditional stochastic resonance Equation (7.39) for sufficient low frequency Ω and for the increasing amplitude A of the input signal [131].

These results have been obtained for non-correlated noise. One can include white noise-like correlations using equations (6.4) with $\kappa \neq 0$. It turns out [133], that for $\kappa \neq 0$, the signal-to-noise ratio depends not only on D, α and κ, but also on the initial condition $x(t = 0)$ of the system. Moreover, for full correlations $\kappa = -1$ and $x(t = 0) = \sqrt{a/b}$ (or $\kappa = 1$ and $x(t = 0) = -\sqrt{a/b}$), the peak becomes very narrow and large ($SNR \approx 10^4$). On the other hand, for $\kappa = -1$ and $x(t = 0) = -\sqrt{a/b}$ (or $\kappa = 1$ and $x(t = 0) = \sqrt{a/b}$), the peak is broaden and becomes very small ($SNR \approx 10^{-7}$).

Another way to complicate the calculations (7.41)-(7.44) is to replace the white multiplicative noise (6.4) by the color Ornstein-Uhlenbeck noise,

$$\langle \xi \left(t_1 \right) \xi \left(t_2 \right) \rangle = \frac{Q}{\tau_1} \exp \left(-\frac{| t_1 - t_2 |}{\tau_1} \right), \qquad (7.45)$$

$$\langle \eta \left(t_1 \right) \eta \left(t_2 \right) \rangle = 2\alpha \delta \left(t_2 - t_1 \right).$$

We omit the cumbersome formulas for $P_{st}(x,t)$, T and SNR obtained in [134], restricting our consideration to the qualitative results. For $\tau_1 = 0$ (white noise) or small τ_1, the signal-to-noise ratio has one peak as a function of the noise strength ratio Q/α, which increases as τ_1 is increased. In addition to this SSR, the second peak appears at a smaller value of Q/α showing the DSR. When $Q/\alpha \to 0$, SNR approaches a finite value, and $SNR \to 0$ when $Q/\alpha \to \infty$. The joint action of κ (correlation between noise) and τ_1 (correlation rate of multiplicative noise) on the signal-to-noise ratio has been considered in [134].

In addition to noise of types (6.4) and (7.45), the more complicate case of white additive noise (7.45) and color multiplicative noise with color coupling between noises

$$\langle \xi(t_1)\xi(t_2)\rangle = \frac{Q}{\tau_1}\exp\left(-\frac{|t_1-t_2|}{\tau_1}\right); \quad \langle \xi(t_1)\eta(t_2)\rangle = \frac{\kappa\sqrt{\alpha Q}}{\tau_2}\exp\left(-\frac{|t_1-t_2|}{\tau_2}\right)$$

$$(7.46)$$

has been considered in [135], where the authors studied the dependence of the signal-to-noise ratio on each of the three parameters (the multiplicative noise correlation time τ_1, the correlation time τ_2 of the coupling between noises and the coupling parameter κ), keeping the other two parameters fixed. It turned out that for all initial conditions, when the value of τ_1 and κ are increased, there a transition appears from one peak to two peaks, and then back to one peak. However, when the value of τ_2 is increased, there appears, depending on initial conditions, a transition either from one peak to two or from two peaks to one, but not two successive transitions. The DSR phenomenon (appearance of two peaks in signal-to-noise ratio as a function of noise strength) can also be induced by dichotomous noise in bistable systems (7.39) [136].

All the preceding examples were related to the simultaneous action of additive and multiplicative noise on a particle moving in a symmetric potential barrier ($U = -\frac{ax^2}{2} + \frac{bx^4}{4}$ in Equation (7.40)). The question arises regarding the form of the stochastic resonance for an asymmetric potential. Two special forms of asymmetric potentials have been considered in [137]. The first is of the form $U = -\frac{x^2}{2} - \frac{ax^3}{3} + \frac{x^4}{4}$, replacing a symmetric potential $U = -\frac{ax^2}{2} + \frac{bx^4}{4}$ in the traditional Equation (7.39), which leads to equation

$$\frac{dx}{dt} = x + ax^2 - x^3 + \eta(t) + A\cos(\Omega t);$$

$$(7.47)$$

the second is of the form $U = \frac{x^4}{4} + \frac{2x^3}{3}$ in the presence of both additive and multiplicative noise, replacing Equation (7.40) by

$$\frac{dx}{dt} = -2x^2 - x^3 + \xi(t)x + \eta(t) + A\cos(\Omega t). \tag{7.48}$$

Analysis of Equation (7.47) shows that the peak of the signal-to-noise ratio as a function of noise strength becomes smaller with increase of a, i.e., the asymmetry of the bistable potential can weaken the phenomenon of stochastic resonance. As for Equation (7.48), it turns out that stochastic resonance occurs not for all values of the parameters in this equation. This phenomenon is absent when there is no multiplicative noise. It is worth mentioning that these results are sensitive to special forms of the asymmetric potential.

An alternative form of an asymmetric potential of the form

$$U(x) = \frac{ax^2}{2} + \frac{bx^4}{4} - hx \tag{7.49}$$

has been considered in [138] which leads to the equation

$$\frac{dx}{dt} = -ax - bx^3 + h + \xi(t)x + \eta(t) \tag{7.50}$$

where $\xi(t)$ and $\eta(t)$ are white noise of strengths D and α, respectively. It turns out [138] that multiplicative noise causes an exponential enhancement of the escape time out of a metastable state. The symmetry breaking field $h \neq 0$ decreases the escape time, although this influence becomes less important for large D and α.

7.6 Mechanism of stochastic resonance

Although the central tenet of stochastic resonance is clear - the non-monotonic dependence of the output signal on the noise strength, the mechanism of this effect is not trivial. Indeed, for too low strength of noise, a particle is not able to overcome the potential barrier, and, contrary to the small decrease of the potential barrier during one half of the period of an external field, jumps over barrier are rare. On the other hand, for too large noise strength, jumps over the barrier are very frequent, and the response of a system is not synchronized with the periodic signal. Therefore, for intermediate noise strength, the transition rate through the barrier is in agreement with the frequency of the input signal. However, the detailed mechanism still remains a mystery: how the input signal is increased due to the presence of noise. It may form a (wrong) impression that the noise transfers part of its energy to the input signal, thereby increasing the output signal. However, another possibility exists which occurs in Nature. Let us recall the transistor, which is composed of three parts – base, collector, and emitter. The collector supplies the large electrical current to the base, and the emitter is the outlet for that supply. This current may be regulated by small current in the emitter. That is the transistor is being used as a switch io increase the current in the collector. As it was shown recently [139], the chaotic signal does not supply energy to a system, but rather plays the role of a switch which allows the amplification of a weak signal. The energetic balance of the underdamped bistable oscillator has been considered [139], in which the dynamics is described by the following equations (with $m = 1$),

$$\frac{dx}{dt} = y; \qquad \frac{dy}{dt} = -\gamma y - \frac{dV}{dx} + f(t) + \xi(t) \qquad (7.51)$$

where $f(t)$ is the periodic force, $V(x)$ is the bistable potential, and $\xi(t)$ is thermal white noise,

$$V(x) = -\frac{1}{2}ax^2 + \frac{1}{4}bx^4; \qquad \langle \xi(t_1)\xi(t_2) \rangle = 2\gamma\kappa T\delta(t_2 - t_1). \qquad (7.52)$$

By using Equation (7.51), one can find the change of the energy of the system, $E = v^2/2 + V(x)$ as

$$\frac{d}{dt}\langle E \rangle = \frac{1}{2}\frac{d}{dt}\langle v^2 \rangle + \frac{d}{dt}\langle V(x) \rangle \qquad (7.53)$$

where $\langle v^2 \rangle$ for the system (7.51) can be found from the solution of the Fokker-Planck equation corresponding to the Langevin Equation (7.51). A simple calculation shows [139] that

$$\frac{d}{dt}\langle E \rangle = -\gamma\langle v^2 \rangle + f(t)\langle v \rangle + \gamma\kappa T \qquad (7.54)$$

The three terms on the right hand side describe, respectively, the energy lost by dissipation, the exchange energy with the driving force and the gain in the thermal energy through the interaction with noise. The last term in (7.54) shows that the power delivered by the noise is constant and does not depend on the amplitude or the frequency of the external force.

Let us average Equation (7.54) over the period T of the external force. Since the internal energy is conserved, $\langle E \rangle_T = 0$, Equation (7.54) leads to

$$\gamma\langle v^2 \rangle_T = \langle f(t)\langle v \rangle \rangle_T + \gamma\kappa T, \qquad (7.55)$$

i.e., the averaged over the period dissipated power consists of the power delivered by the external force plus the constant power delivered by the noise. It is remarkable that the power delivered by the noise is constant and, therefore, does not influenced by the parameters of an input signal. In other words, noise does not support the energy for the amplification of the input signal, but rather, plays the role of a tuner, helping the system to absorb more energy from the external force. These conclusions have been supported [139] by numerical solutions of the nonlinear Equation (7.51) with subsequent averages.

Thus, the role of noise in the stochastic resonance phenomenon is quite analogous to the role of the emitter current in a transistor, described in the beginning of this section.

7.7 Fluctuating potential barrier (resonance activation)

In Chapter 6 we considered the stationary properties of metastable states subject to fluctuations of the potential barriers which leads to noise-enhanced stability. Here we consider a similar phenomena for a bistable potential which will result in a maximum of the escape rate at some "resonant" fluctuation rate (resonant activation). The distinction between resonance activation and the stochastic resonance considered in the previous section lies in the fact that modulation of a potential barrier is random in the former case but deterministic in the latter case.

7.7.1 *Piece-wise linear potential*

Just as a simplified piece-wise squared potential has been used in the previous chapter to obtain the exact analytical solution, one can use a piece-wise linear potential for the description of the resonance activation. This potential is shown in Figure 7.2 under the assumption that the potential increases to infinity at $x = \pm L$ and fluctuates randomly between two configurations $V_{\pm}(x)$ with flipping rate λ (dichotomous noise). Such a model has been analyzed in [72],

The mean first-passage time T is considered for a particle subject to white noise, starting at the bottom of a potential well at $x = L$ in Figure 7.2, and reaching the top of the barrier at $x = 0$. The boundary conditions are assumed to be reflecting at $x = L$ and absorbing at $x = -L$. Using equations (7.34)-(7.35), for the case where the midpoint of the barriers fluctuates between $\pm E$, one obtains in terms of the dimensionless variables

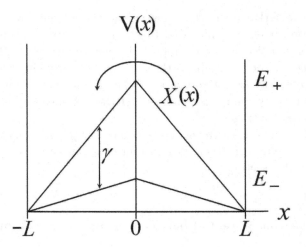

Fig. 7.2   The triangle piece-wise potential subject to the dichotomous fluctuations of the barrier height. [Reprinted figure with permission from Ref. 72. Copyright (1992) by the American Physical Society.]

$\tau = \frac{TD}{L^2}$ and $\varepsilon = \frac{\lambda L^2}{D}$ [72]

$$\tau = A_- \left[-\frac{E}{Dk}(1 - \exp(k)) + \frac{\varepsilon Dk}{E} - \frac{E}{D} \right]$$

$$+ A_+ \left[\frac{E}{Dk}(1 - \exp(-k)) - \frac{\varepsilon Dk}{E} - \frac{E}{D} \right] \tag{7.56}$$

where

$$A_\pm = \frac{2\varepsilon \left[\pm \exp\left(\pm k\right) - k \right] \pm \frac{E^2}{D^2}}{\frac{2Ek}{D} \left\{ 1 + \frac{2\varepsilon D^2}{E^2} \left[2\varepsilon \cosh\left(k\right) + \frac{E^2}{D^2} \right] \right\}}; \quad k = \sqrt{\frac{E^2}{D^2} + 2\varepsilon}. \tag{7.57}$$

Analysis of Equation (7.56), supported by the Monte Carlo simulations, shows [72] that the mean free-passage time is a non-monotonic function of the fluctuation rate, reaching minimum at a "resonant" rate (resonant activation).

A quite general form of the potential was considered in [140], making only the assumption that the potential barrier is higher than the noise strength. It turns out that resonant activation is a typical phenomenon which has a simple physical interpretation. For a very fast and a very slow fluctuation rate, the escape rate (reciprocal mean free-passage time) is determined by the average barrier height and by the highest barrier height, respectively. For the intermediate regime, the rate is given by the average rate which is greater than the rate for fast or slow fluctuations, which assures the maximum of the escape rate.

An interesting phenomenon occurs when, in addition to additive white noise considered in [72], [140], another dichotomous additive noise acts on the particle crossing a fluctuating barrier. It turns out [141] that the mean free-passage time displays two resonant activations: one (as described above) as a function of the flipping rate of the fluctuating potential barrier, and the other as a function of the correlation rate of dichotomous noise. Dichotomous noise can weaken the former resonant activation but enhance the latter one.

7.7.2 *Phenomenological model*

The general feature of stochastic activation is supported not only by the complication of the original model [72] made in [141], but also by its simplification. Resonant activation has been obtained [142] for a particle randomly switched at a rate λ between two states called $+$ and $-$, respectively. In addition, a particle may leave each of these states at time t with probabilities $\psi_+ (t)$ and $\psi_- (t)$. Three different forms of these functions have been considered in [142].

a) The particles switch between two states with characteristic rates k_+ and k_-. For this case, the mean free-passage time is equal to

$$T = \frac{(k_+ + k_-)/2 + 2\lambda}{k_+ k_- + \lambda (k_+ + k_-)} \tag{7.58}$$

which is a monotonic function of λ decreasing from $\frac{1}{2} (1/k_+ + 1/k_-)$ for $\lambda = 0$ to $2/(k_+ + k_-)$ for $\lambda = \infty$, i.e., there is no resonant activation.

b) The probability densities for leaving the states are $\psi_+ (t) = \delta \left(t - k_+^{-1} \right)$ and $\psi_- (t) = \delta \left(t - k_-^{-1} \right)$. For this case,

$$T = \frac{1 + 2 \exp \left[\lambda \left(1/k_+ + 1/k_- \right) \right] - 3/2 \left[\exp \left(\lambda/k_+ \right) + \exp \left(\lambda/k_- \right) \right]}{\lambda \left[\exp \left(\lambda/k_+ \right) + \exp \left(\gamma/k_- \right) - 1 \right]}. \tag{7.59}$$

It follows from Equation (7.59) that for $\lambda \to 0$, $T \to \left(k_+^{-1} + k_-^{-1} \right)/2$, and that T diverges for $\lambda \to \infty$. However, T goes through a minimum when k_+ and k_- differ by at least 3.75, showing thereby the resonant activation.

c) For the case of $\psi_+ (t) = \psi_- (t)$, resonant activation exists for all forms of this function.

7.7.3 Coherent stochastic resonance

In addition to the simple model described in the previous section, there is another simple linear system which behaves similarly regarding resonant activation. This is one-dimensional diffusion on a segment terminated by one or two traps with a periodic force of small amplitude so that the particle cannot reach a trap in the absence of noise [143]. Although the system remains linear, the trapping boundaries produce additional states and the particle can be trapped or untrapped. The mean free-passage time to trapping has been shown to vary non-monotonically as a function of both the frequency [143] and the amplitude [144] of the periodic force (coherent stochastic resonance). An external periodic force supplies the characteristic frequency just like the flipping rate between two positions of the potential barriers in the case of resonant activation. Still another possibility for introducing the characteristic frequency was considered in [145], where a random walk on a finite interval terminated by one or two traps occurred in such a way that at any time, each site has one or two transition rates changing at random. Such system also shows coherent stochastic resonance. The explanation of this phenomenon is similar to that for resonant activation, and connected with the different behavior of the mean free-passage time at low and high frequencies of a periodic field [146]. At very low frequency, $\sin(\Omega t) \approx \Omega t$, and when the frequency increases from zero, T necessarily decreases, becoming smaller than its value at $\Omega = 0$. On the other hand, for $\Omega \to \infty$, the rate of oscillation increases, the particle is less and less influenced by the periodic force, and T approaches its value of a pure diffusive process as it was the case for $\Omega = 0$. It is clear, therefore, that T exhibits a minimum at some frequency. If a periodic force is replaced by a periodic [146] or random [147] telegraph signal, the different behavior at small frequency eliminates coherent stochastic resonance.

Chapter 8

Harmonic Oscillator with Additive Noise

8.1 Internal and external noise

The equation of motion for $x(t)$ has a different form for external and internal noise. For external noise, the friction coefficient γ and the correlation function of the noise are independent, and the equation of motion has the following form

$$\frac{d^2x}{dt^2} + 2\gamma\frac{dx}{dt} + \omega^2 x = \xi(t) \tag{8.1}$$

while for internal noise, the dissipation and noise stem from the same source, and the system will finally reach an equilibrium state. Then, equation (8.1) has to be rewritten as

$$\frac{d^2x}{dt^2} + 2\int_0^t \gamma(t-t_1)\frac{dx}{dt}(t_1)\,dt_1 + \omega^2 x = \xi(t) \tag{8.2}$$

where, due to the fluctuation-dissipation theorem,

$$\langle \xi(t)\xi(t+\tau)\rangle = 2\kappa T\gamma(\tau). \tag{8.3}$$

8.2 White and dichotomous noise

After averaging equation (8.1), one immediately finds for the first moment

$$\langle x(t)\rangle = \exp(-\gamma t)\left[\frac{\gamma x(t=0) + \frac{dx}{dt}(t=0)}{\omega_1}\sin(\omega_1 t) + x(t=0)\cos(\omega_1 t)\right], \tag{8.4}$$

where $\omega_1 = \sqrt{\omega^2 - \gamma^2}$.

The stationary distribution function $P_{st}(x)$ can easily be found from the Fokker-Planck equation. For the equilibrium system (8.2), $P_{st}(x)$ has the canonical equilibrium distribution,

$$P_{st}\left(x, \frac{dx}{dt}\right) = N \exp\left(-\frac{\omega^2 x^2}{2\kappa T} - \frac{(dx/dt)^2}{2\kappa T}\right) \tag{8.5}$$

which leads to the following limit values as $t \to \infty$,

$$\left\langle [x(t) - \langle x(t)\rangle]^2 \right\rangle = \frac{\kappa T}{\omega^2}; \quad \left\langle [v(t) - \langle v(t)\rangle]^2 \right\rangle = \kappa T. \tag{8.6}$$

For external noise, the form of $P_{st}(x)$, which corresponds to equation (8.1), depends on the properties of the noise. For Ornstein-Uhlenbeck noise (2.6), it is convenient [148] to express $P_{st}(x)$ in terms of the following variable

$$q = \varepsilon(t) - 2\gamma \frac{dx}{dt} - \frac{2\gamma\tau\omega^2}{1 + 2\gamma\tau}x \tag{8.7}$$

where $\varepsilon(t)$ is the white noise, $\langle \varepsilon(t_1)\varepsilon(t_2)\rangle = 2\delta(t_1 - t_2)$.

Then, one gets,

$$P_{st}\left(x, \frac{dx}{dt}, q\right) = N \exp -\frac{1}{D}\left[\gamma\omega^2\left(1 + \frac{\omega^2\tau^2}{1 + 2\gamma\tau}\right)x^2 +\right] +$$

$$\gamma\left(1 + 2\gamma\tau + \omega^2\tau^2\right)\left(\frac{dx}{dt}\right)^2 + \frac{\tau(1 + 2\gamma\tau)}{2}q^2. \tag{8.8}$$

The stationary second moment can easily be obtained from (8.8). For an exponential correlation function (2.6), the variance turns out [149] to be equal to

$$\left\langle [x(t) - \langle x(t)\rangle]^2 \right\rangle = \frac{2D\tau}{1 - 2\gamma\tau + \omega^2\tau^2}\left\{\frac{\tau}{1 + 2\gamma\tau + \omega^2\tau^2}\left[1 - \right.\right.$$

$$\left(\frac{(1 + 2\gamma\tau)\sin\omega_1 t}{2\omega_1\tau} + \cos\omega_1 t\right)\exp\left(-\frac{1 + 2\gamma\tau}{2\tau}t\right)\right] + \frac{1 - 2\gamma\tau}{4\omega^2\gamma\tau}[1 - \exp(-2\gamma t)] -$$

$$\frac{\exp(-2\gamma t)}{4\omega_1^2}\left[\frac{1 - 2\gamma\tau}{\omega_1\tau}\sin 2\omega_1 t + \frac{2\gamma - 4\gamma^2\tau + 2\omega^2\tau}{\omega_1^2\tau}\sin^2\omega_1 t\right]\right\}. \tag{8.9}$$

For white noise, this equation reduces to

$$\left\langle [x(t) - \langle x(t)\rangle]^2 \right\rangle = \frac{D}{2\gamma\omega^2}\left\{1 - \left[1 + \frac{\gamma}{\omega_1}\left(\sin 2\omega_1 t + \frac{2\gamma}{\omega_1}\sin^2\omega_1 t\right)\right]\exp(-2\gamma t)\right\}. \tag{8.10}$$

The stationary $(t \to \infty)$ values of the variance are the following: for color noise,

$$\lim_{t\to\infty}\left\langle [x(t) - \langle x(t)\rangle]^2 \right\rangle = \frac{D(1 + 2\gamma\tau)}{2\omega^2\gamma(1 + 2\gamma\tau + \omega^2\tau^2)} \tag{8.11}$$

and for white noise,

$$\lim_{t \to \infty} \left\langle [x(t) - \langle x(t) \rangle]^2 \right\rangle = \frac{D}{2\omega^2\gamma}. \qquad (8.12)$$

For the following discussion, let us note that the stationary value (8.11) of the variance turns out to be a non-monotonic function of the correlation rate τ^{-1}.

Both equations (8.10) and (8.11) were already obtained in 1945 [150].

It turns out [149], that the mean-squared coordinate and velocity show ballistic behavior at small times, $t \to 0$: for color noise, both internal and external,

$$\left\langle (\Delta x)^2 \right\rangle \sim t^4 \quad \text{and} \quad \left\langle (\Delta v)^2 \right\rangle \sim t^2 \qquad (8.13)$$

while for white noise,

$$\left\langle (\Delta x)^2 \right\rangle \sim t^3 \quad \text{and} \quad \left\langle (\Delta v)^2 \right\rangle \sim t. \qquad (8.14)$$

One can also study [149] the change in the time-dependence of the variance for different time intervals depending on the relation between the observation time t and the three characteristic times of the problem: the relaxation time $1/\gamma$, the period of the force-free oscillator $2\pi/\omega$, and the correlation time τ.

8.3 Additive noise and parametric oscillations

In previous sections we considered linear differential equations with constant coefficients. However, in many practical applications these coefficients are periodic functions of time leading, in particular, to the so-called Mathieu equation .

$$\frac{d^2 x}{dt^2} + (a - 2b\cos t)x = \xi(t) \qquad (8.15)$$

The properties of the dynamic Mathieu equation ($\xi(t) = 0$ in (8.15)) for different values of a and b are well known [151][1].

Solution of equation (8.15) with random noise $\xi(t)$ has basically the same qualitative properties as the deterministic equation. The first moment

---

[1]Equation (8.15) can even be applied to the dynamic behavior of an acrobat who holds an assistant poised on a pole above his head while he himself stands on a spherical boll rolling on the ground. The analysis of this system, however, requiers further experimental investigation, which we leave to the interesting reader.

$\langle x \rangle$ remains unchanged while for white noise one gets [152] the following equation for the variance $\sigma^2 \equiv \langle x^2 \rangle - \langle x \rangle^2$

$$\frac{d^3 \sigma^2}{dt^3} + 4 \left[a - 2b \cos\left(2t \right) \right] \frac{d\sigma^2}{dt} + 8b \sin\left(2t \right) \sigma^2 = 2D. \qquad (8.16)$$

As it follows from (8.16), the variance σ^2 increases exponentially with time in place of the more usual linear increase in the deterministic case.

Chapter 9

Nonlinear Oscillator with Additive Noise

9.1 Statistical linearization

The first attempt to solve a nonlinear problem is to replace it (accurately as possible) by a linear problem which allows an exact solution. For stochastic differential equation,

$$\frac{d^2 x}{dt^2} + \gamma \frac{dx}{dt} + \omega_0^2 x + bf(x) = \sqrt{2D}\xi(t), \tag{9.1}$$

where $bf(x)$ is a nonlinear force, we replace Equation (9.1) by the "equivalent" linear equation

$$\frac{d^2 x}{dt^2} + \gamma \frac{dx}{dt} + \omega^2 x = \sqrt{2D}\xi(t) \tag{9.2}$$

in such a way that the error

$$\Delta(x) = \omega_0^2 x + bf(x) - \omega^2 x \tag{9.3}$$

is minimal. In the method of statistical linearization [153] ω is chosen to minimize the mean square error $\langle \Delta^2(x) \rangle$ at equilibrium. Thus, ω is the solution of the equation

$$\lim_{t \to \infty} \frac{\partial}{\partial \omega^2} \langle \Delta^2(x) \rangle \equiv \frac{\partial}{\partial \omega^2} \langle \Delta^2(x) \rangle_{as} = 0 \tag{9.4}$$

where the average is performed over all realizations of the random force $\xi(t)$ at equilibrium,

$$\langle \Delta^2(x) \rangle_{as} = \int_{-\infty}^{\infty} \Delta^2(x) P_{as}(x) \, dx. \tag{9.5}$$

It was shown [154] that the procedure (9.4) gives the best approximation for the first two moments, namely, the first asymptotic moment $\langle x(t) \rangle_{as}$

is the same for (9.1) and (9.2), while the difference between the second moments $\langle x^2(t) \rangle_{as}$ is minimized.

On substitution (9.3) into (9.4), one obtains

$$\omega^2 = \omega_0^2 + b \frac{\langle x f(x) \rangle_{as}}{\langle x^2 \rangle_{as}}. \tag{9.6}$$

Then, accordingly to (9.5)-(9.6), the problem is reduced to finding the asymptotic probability distribution $P_{as}(x)$ for the linear Equation (9.2). For Gaussian white noise and zero initial conditions,

$$P_{st}(x) = \frac{1}{\sqrt{2\pi\sigma^2}} \exp\left(-\frac{x^2}{2\sigma^2}\right) \tag{9.7}$$

with $\sigma^2 = \langle x^2 \rangle = D/\gamma\omega^2$. If the non-linear function $f(x)$ is odd in x,

$$f(x) = \sum_{n=1}^{\infty} f_n x^{2n+1} \tag{9.8}$$

then [154]

$$\frac{D}{\gamma\sigma^2} = \omega_0^2 + b \sum_{n=1}^{\infty} f_n c_n \sigma^{2n} \tag{9.9}$$

where the coefficients c_n are given by

$$c_n = \int_{-\infty}^{\infty} x^{2n+2} \exp\left(-\frac{x^2}{2}\right) dx. \tag{9.10}$$

On substituting (9.10) into (9.9) one obtains the equation for the variance σ^2 which generally, has to be solved numerically. The corrections to the outlined process of statistical linearization have been considered in [154].

9.2   Double-well oscillator with additive noise

The motion of a particle in a double-well potential $U = -\frac{\omega^2 x^2}{2} + \frac{bx^4}{4}$ is described by the following equation

$$\frac{d^2 x}{dt^2} + 2\gamma \frac{dx}{dt} = -\frac{dU}{dx} + \sqrt{2D}\xi(t) = \omega^2 x - bx^3 + \sqrt{2D}\xi(t) \tag{9.11}$$

which is a nonlinear generalization of (8.1).

A detailed analysis has been carried out [155] of the correlation function $\langle x(t) x(0) \rangle$ and its Fourier component $Q(\Omega)$ (power spectrum) defined as

$$Q(\Omega) = \frac{1}{\pi} \text{Re} \int_0^{\infty} dt \, \exp(\Omega t) \langle x(t) x(0) \rangle, \tag{9.12}$$

In terms of the dimensionless "time" ωt and "coordinate" $\frac{\sqrt{b}}{\omega}x$, two characteristic dimensionless parameters are γ/ω and β, where

$$\beta = \frac{bD}{2\omega^4} = \frac{D}{8\Delta U} \qquad (9.13)$$

with ΔU being the height of the potential barrier, $\Delta U = \omega^4/4b$.

It turns out [155] that for the underdamped oscillator, $\gamma/\omega \ll 1$, the function $Q(\Omega)$ has either two or three distinct peaks, depending on the noise intensity β. One of the peaks is located at frequency $\Omega = 0$. For $\beta \ll 1$, this peak is related to the fluctuational transitions between the two stable points of the potential. The intensity of this peak is of order of ω^2/b, and its half-width is exponentially small, being proportional to $\gamma \exp\left[-(4\beta)^{-1}\right]$. Another peak at $\beta \ll 1$ is related to the small amplitude intrawell vibrations, and this peak is located at $\Omega \simeq \sqrt{2}\omega$. With increasing noise intensity, an increasingly important role is played by the third peak related to the overbarrier vibrations and located at $\Omega \simeq \pi\omega/\ln\left(16\beta^{-1}\right)$ for $\ln\left(16\beta^{-1}\right) \gg 1$. The detailed calculations of shapes, intensities and locations of the peaks have been supported by analogue experiments using an electronic circuit [155].

9.3 Double-well oscillator driven by two periodic fields (vibrational resonance)

As we have already seen in Chapter 5, a shift of the stable points occurs either through multiplicative noise or through a parametric periodic force even though the physical mechanism is quite different. Indeed, this effect is determined by the low frequencies of the correlator of noise, and by the high frequencies of an external field. It turns out [156] that the analogous effect occurs for two additive periodic fields as well. A bistable underdamped oscillator subject to two periodic fields is described by the following equation

$$\frac{d^2x}{dt^2} + \gamma\frac{dx}{dt} - \omega_0^2x + bx^3 = A\sin(\omega t) + C\sin(\Omega t). \qquad (9.14)$$

Suppose now that one of the fields has large amplitude (larger than the barrier height $\Delta U = \omega_0^2/4b$), $C > \omega_0^2/4b$, and high frequency, $\Omega \gg \omega$. The former means that during each half-period, this field transfers the system from one potential well to the other. A similar situation holds in a random system where the large amplitude field in (9.14) is replaced by an additive random force, which plays the same role of switching a system between the two minima. Therefore, by choosing the appropriate relation between the input signal $A \sin(\omega t)$ and the amplitude C of the large signal (or the strength of noise), one can obtain a non-monotonic dependence of the output signal on the noise strength (stochastic resonance considered in section 5.3) or on the amplitude C (vibrational resonance [156]).

For the qualitative description of vibrational resonance, consider Equation (9.14) [102]. Due to the high frequency Ω, one has two different time scales. Therefore, let us seek a solution of Equation (9.14) in the form

$$x(t) = y(t) - \frac{C \sin(\Omega t)}{\Omega^2}. \tag{9.15}$$

The first term on the right-hand side will be assumed to vary significantly only over times of t, while the second one varies rapidly. On substituting (9.15) into (9.14), one can average over a single cycle time of $\sin(\Omega t)$. All odd powers of $\sin(\Omega t)$ vanish upon averaging, while the $\sin^2(\Omega t)$ term gives $1/2$. Finally, one obtains the following equation for $X(t)$, the mean value of $y(t)$, $X(t) = \langle y(t) \rangle$,

$$\frac{d^2 X}{dt^2} + \gamma \frac{dX}{dt} - \left(\omega_0^2 - \frac{3bC^2}{2\Omega^4} \right) X + bX^3 = A \sin(\omega t) \tag{9.16}$$

where the slowly varying term $A \sin(\omega t)$ does not change under averaging over short times.

The approximate solution of Equation (9.16) will be given in Chapter 14. Here we perform a quantitative analysis of this Equation [157]. Equation (9.16) describes the driven motion in the effective potential U_{eff} of the form

$$U_{\text{eff}} = \left(\omega_0^2 - \frac{3bC^2}{2\Omega^4} \right) \frac{X^2}{2} - \frac{bX^4}{4}. \tag{9.17}$$

Hence, the locations of new equilibrium states about which slow oscillations are executed, depend on the parameter $3bC^2/2\omega_0^2\Omega^4$. If this parameter is smaller than unity, $\left(3bC^2/2\omega_0^2\Omega^4 \right) < 1$, there are two equilibrium states

$$X_{1,2} = \pm \sqrt{\frac{\omega_0^2}{b} - \frac{3C^2}{\Omega^4}}. \tag{9.18}$$

For $\left(3bC^2/2\omega_0^2\Omega^4 \right) > 1$, there is only one equilibrium state, $X = 0$. Thus, transition from bimodal to unimodal distribution depends on the amplitude of the high-frequency field (analogous to the shift of the stable points induced by noise, considered in section 5.1).

The equation for the deviation, $Y = X - X_{1,2}$, of X from one of the stable equilibrium states $X_{1,2}$, is obtained by the substitution $X = Y + X_{1,2}$ into Equation (9.16),

$$\frac{d^2Y}{dt^2} + \gamma\frac{dY}{dt} + 2\left(\omega_0^2 - \frac{3bC^2}{2\Omega^4}\right)Y + 3bX_{1,2}Y^2 + bY^3 = A\sin\left(\omega t\right), \quad (9.19)$$

while the derivation from $X = 0$ is described by (9.16). We see that the resonance frequency of the system

$$\omega_{res,1} = \sqrt{2\left(\omega_0^2 - \frac{3bC^2}{2\Omega^4}\right)} \qquad (9.20)$$

decreases from $\sqrt{2}\omega_0$ to zero as the amplitude C increases from zero to $C_0 = \sqrt{2\omega_0^2\Omega^4/3b}$.

On the other hand, for $X = 0$, the resonance frequency

$$\omega_{res,2} = \sqrt{\frac{3bC^2}{2\Omega^4} - \omega_0^2} \qquad (9.21)$$

increases with a decrease in C from C_0. Therefore, one concludes that the resonance frequency is non-monotonic function of C. Numerical solutions of linearized Equations (9.19) for $C < C_0$ and (9.16) for $C > C_0$, show [157] that the amplitude of the output signal has a maximum as a function of the amplitude C of the high-frequency field. Moreover, the *bona fide* resonance occurs as a function of the high frequency Ω when the low frequency ω is fixed, as well as being a function of ω at fixed Ω.

All the above results have been obtained for an underdamped oscillator. It turns out [157] that similar effects also take place for an overdamped oscillator. The similar calculation has been performed recently, [158], analyzing the role of depth and location of minima of a double-well potential on vibration resonance in both underdamped and overdamped Duffing oscillator. While for the underdamped system one or two resonances are possible, depending on the depth and location of the minima of the potential well, it turns out that in the overdamped system only one resonance is possible. The value of parameter C in Equation (9.14) at which resonance occur, independent on the depth of the wells of the symmetric double-well potential, but varies linearly with the location of the minima of the wells.

Experimental evidence for vibrational resonance has been obtained in an optical system [159] and in an electronic circuit [160].

9.4 Stochastic resonance in linear system subject to two periodic fields and random forces

Thus far, we considered the resonance action of a periodic force on a system with random parameters (stochastic resonance) or on a system subject to two periodic forces (vibrational resonance). The dynamic equation, which combines these two cases, has the following form,

$$\frac{dx}{dt} = -ax^3 + r + \xi(t) x + \eta(t) + f_1 \cos(\Omega_1 i) + f_2 \cos(\Omega_2 t) \qquad (9.22)$$

where r is the constant bias of the monostable system, $\xi(t)$, and $\eta(t)$ are white noises of intensities D and Q, respectively. Analysis of the stationary distribution function of the Fokker-Planck equation corresponded to the Langevin Equation (9.22) shows [161] that the monostable system (9.22) is equivalent to the bistable system with stable points $x_\pm = \pm\sqrt{D/a}$ and unstable point $x_0 = 0$. We don't bring here the cumbersome formulas for the signal-to-noise ratio, which shows the resonance dependence on the parameters a, r and multiplicative noise intensity D [161]. While both dependence of SNR as functions of D and Q are non-monotonic, the positions of maxima move to the right with increasing r in the former case, and to the left in the latter case. Another peculiarity is the availability of the resonance as a function of r, which is absent in the case of an asymmetric bistable system.

Chapter 10

Harmonic Oscillator with Random Frequency

10.1 First moment for the random frequency

10.1.1 *Force-free oscillator*

The equation of motion of the oscillator with a random frequency

$$\frac{d^2x}{dt^2} + 2\gamma\frac{dx}{dt} + \omega^2\left[1 + \xi(t)\right]x = 0 \qquad (10.1)$$

has been studied extensively (The comprehensive list of references can be found in [162]). We list here the main results for white and color noise.

10.1.2 *White noise*

It turns out that the fluctuations of frequency do not effect the first moment of the underdamped oscillator provided the fluctuations are delta correlated, and $\langle x(t) \rangle$ remains equal to noise-free solution $x(t)$ defined in (1.4).

10.1.3 *Color noise*

To get the equation for the averaged first moment from Equation (10.1) with color noise, one has to use some approximate methods. The approximations are different for slow and fast fluctuations where the correlation rate is much smaller or larger than the characteristic time of the deterministic system. For the slow (adiabatic) fluctuations, one uses an expansion in small parameter $\sigma^2/\omega_0 < 1$ which leads to [163]

$$\frac{d^2 \langle x \rangle}{dt^2} + 2\gamma \left(1 - \langle \xi^2 \rangle\right) \frac{d \langle x \rangle}{dt} +$$

$$\left(\omega^2 + \omega^2 \langle \xi^2 \rangle + 2\gamma \left\langle \xi \frac{d\xi}{dt} \right\rangle\right) \langle x \rangle = 0. \tag{10.2}$$

For the fast fluctuations, the small parameter is $\sigma^2 \tau < 1$, and the equation is of the form [34]

$$\frac{d^2 \langle x \rangle}{dt^2} + \left(2\gamma + \omega^2 c_2\right) \frac{d \langle x \rangle}{dt} + \omega^2 \left(1 - \omega c_1\right) \langle x \rangle = 0 \tag{10.3}$$

where the coefficients

$$c_1 = 2 \int_0^\infty \langle \xi\left(t\right) \xi\left(t - \tau\right)\rangle \sin\left(2\omega\tau\right) d\tau, \tag{10.4}$$

$$c_2 = 2 \int_0^\infty \langle \xi\left(t\right) \xi\left(t - \tau\right)\rangle \left[1 - \cos\left(2\omega\tau\right)\right] d\tau.$$

In agreement with the foregoing these coefficients vanish for the white noise correlators. Equations (10.2) and (10.3) show that the fluctuations in frequency cause a gain in $\langle x \rangle$ for adiabatic fluctuations and damping for fast fluctuations.

Stability conditions of Equations (10.1) have been considered [164] also for a signal-modulated color noise with the correlator (2.7). It turned out [164] that the maximum stability of power spectrum $S\left(\Omega\right)$ occurs at "resonance" frequencies $\Omega = 2\omega_2/n$. Numerical analysis shows that for large correlation rate λ, the main resonance $(n = 1)$ appears at $\Omega = 2\omega_2$, while for small λ, one sees additional resonances $(n = 2, 3)$ at $\Omega = \omega_2$ and $\Omega = 2\omega_2/3$. Similar maxima arise for Equation (10.1) with random force $\xi\left(t\right)$ replaced by the periodic force $A \cos\left(\omega_2 t\right)$.

10.2   Second moment for a random frequency

The first moment for an oscillator with random frequency in the presence of an external periodic field was calculated in Chapter 5. In the absence of an external field there are different ways to calculate the second moment of Equation (10.1). We bring here without proof only few results. Firstly, the second moment of the oscillator with random frequency subjected to dichotomous multiplicative noise $\xi(t)$ of strength σ^2 and correlation rate λ, and additive white noise $\eta(t)$ defined as $\langle \eta(t_1) \eta(t_2) \rangle = 2\alpha\delta(t_1 - t_2)$ is given by [165]

$$\langle x^2 \rangle = 2\alpha \left[4\gamma\omega^2 - \frac{2\omega^4\sigma^2(\lambda + 4\gamma)^2}{(\lambda + 2\gamma)[4\omega^2 + \lambda(\lambda + 4\gamma)]} \right]^{-1} \tag{10.5}$$

which for the white noise $\xi(t)$ reduces to

$$\langle x^2 \rangle = \frac{\alpha}{\omega^2(2\gamma - D\omega^2)} \tag{10.6}$$

showing the well known "energetic" instability for $D\omega^2 > 2\gamma$.

Analogous calculations for equation

$$\frac{d^2x}{dt^2} + \gamma\frac{dx}{dt} + \omega^2[1 + \xi(t)]x = \eta(x) \tag{10.7}$$

with white noise $\eta(x)$, $\langle \eta(x_1)\eta(x_2) \rangle = 2\alpha\delta(x_1 - x_2)$, and Poisson noise $\xi(t)$ defined by Poisson parameter λ and jump length S, gives for a second moment [166]

$$\langle x^2 \rangle = \frac{\alpha\omega^2}{\gamma\omega^2 - \lambda S^2/2}, \tag{10.8}$$

i.e., the instability occurs at $\lambda S^2 > 2\gamma\omega^2$.

Another way to find the first and second moments is the use of the Laplace transformation

$$\langle x(p) \rangle = \int_0^\infty \langle x(t) \rangle \exp(-pt)\, dt. \tag{10.9}$$

Then, one obtains [13] for $\langle x \rangle$ with $\gamma = 0$ and initial conditions $x(t = 0) = x_0$, $\frac{dx}{dt}(x = 0) = y_0$,

$$\langle x(p) \rangle = \frac{[px_0 + y_0]\left[(p + \lambda)^2 + \omega^2\right]}{(p^2 + \omega^2)\left[(p + \lambda)^2 + \omega^2\right] - \omega^4\sigma^2}, \tag{10.10}$$

and for $\langle x^2(p) \rangle$ and $x(t = 0) = 0$, $\frac{dx}{dt}(x = 0) = y_0$,

$$\langle x^2(p) \rangle = \frac{2y_0^2 p\left[(p + \lambda)^2 + \omega^2\right]}{p^2(p^2 + \omega^2)\left[(p + \lambda)^2 + \omega^2\right] - 16\omega^4\sigma^2(2p + \lambda)^2}. \tag{10.11}$$

10.3   Maxwell equation with random dielectric constant

We replace Equation (10.1) with $\gamma = 0$ by the isomorphic Maxwell equation with an external force, substituting the electric field E and coordinate x instead of the coordinate x and time t, and adding the periodic field $a \sin (kx)$,

$$\frac{d^2 E}{dx^2} + [\varepsilon_0 + \varepsilon_1 \xi (x)] E = a \sin (kx) \tag{10.12}$$

where for white noise

$$\langle \xi (x) \xi (x_1) \rangle = D\delta (x - x_1) \tag{10.13}$$

and for dichotomous noise

$$\langle \xi (x) \xi (x_1) \rangle = \sigma^2 \exp (-\lambda |x - x_1|) . \tag{10.14}$$

There are also other models, isomorphic to that described by Equation (10.1), such as the theory of paramagnetic resonance, high-dimensional Hamiltonian systems [167], and Anderson localization in solids for which Equation (10.1) is the one-dimensional Shrodinger equation for a single particle in a delta-correlated potential $\xi (t)$ [168].

The second-order differential Equation (10.12) can be rewritten as two first-order differential equations,

$$\frac{dE}{dx} = y, \tag{10.15}$$

$$\frac{dy}{dx} = -\varepsilon_0 E - \varepsilon_1 \xi (x) E + a \sin (kx) .$$

The subsequent calculations are similar to those performed for harmonic oscillator and described in the next section, Equation (10.15) is similar to (10.28)-(10.29). The calculations quite similar to (10.30)-(10.38), lead to the following fourth-order differential equation for $\langle E \rangle$

$$\frac{d^4 \langle E \rangle}{dx^4} + 2\lambda \frac{d^3 \langle E \rangle}{dx^3} + \left(2\varepsilon_0 + \lambda^2\right) \frac{d^2 \langle E \rangle}{dx^2} + 2\varepsilon_0 \lambda \frac{d \langle E \rangle}{dx} + \tag{10.16}$$
$$\left[\varepsilon_0 \left(\varepsilon_0 + \lambda^2\right) - \varepsilon_1 \sigma^2\right] \langle E \rangle = \left(\varepsilon_0 + \lambda^2 - k^2\right) a \sin kx + 2\lambda ak \cos kx.$$

The response to an external field is defined by the solution of the non-homogeneous Equation (10.16) which has the following form

$$\langle E \rangle_A = A \sin (kx + \varphi) \tag{10.17}$$

with

$$A = a \left[\frac{\left[\left(k^2 - \varepsilon_0 - \lambda^2 \right)^2 + 4\lambda^2 k^2 \right]}{\left[\left(k^2 - \varepsilon_0 \right) \left(k^2 - \varepsilon_0 - \lambda^2 \right) - \varepsilon_1^2 \sigma^2 \right]^2 + 4\lambda^2 k^2 \left(k^2 - \varepsilon_0 \right)^2} \right]^{\frac{1}{2}}, \quad (10.18)$$

$$\tan \varphi = \frac{2k\lambda \varepsilon_1 \sigma^2}{\left(k^2 - \varepsilon_0 \right) \left[\left(k^2 - \varepsilon_0 - \lambda^2 \right)^2 + 4k^2 \lambda^2 \right] - \left(k^2 - \varepsilon_0 - \lambda^2 \right) \varepsilon_1^2 \sigma^2}.$$

Analogously, one obtains

$$\langle y \rangle_A = Ak \cos(kx + \phi); \quad \langle \xi E \rangle_A = L \sin(kx + \phi) + M \cos(kx + \phi); \quad (10.19)$$

$$\langle \xi y \rangle_A = (\lambda L - kM) \sin(kx + \phi) + (\lambda M + kL) \cos(kx + \phi)$$

where

$$L = \frac{k^2 - \varepsilon_0 - \lambda^2}{2\lambda k} M, \qquad M = \frac{2Ak\lambda \varepsilon_1^2 \sigma^2}{\left[\left(k^2 - \varepsilon_0 - \lambda^2 \right)^2 + 4k^2 \lambda^2 \right]}. \quad (10.20)$$

To find the second moments, one multiplies Equations (10.15) by $2E$ and $2y$, respectively, which gives

$$\frac{d}{dx} \langle E^2 \rangle = 2 \langle Ey \rangle \quad (10.21)$$

$$\frac{d}{dx} \langle y^2 \rangle = -2\varepsilon_0 \langle Ey \rangle - 2\varepsilon_1 \langle \xi Ey \rangle + 2a \langle y \rangle \sin kx.$$

Analogously, multiplying Equations (10.15) by y and E, and summarizing these equations and averaging, results in

$$\frac{d}{dx} \langle Ey \rangle = \langle y^2 \rangle - \varepsilon_0 \langle E^2 \rangle - \varepsilon_1 \langle E^2 \xi \rangle + a \langle E \rangle \sin kx. \quad (10.22)$$

Equation (10.22) contains a new correlator $\langle E^2 \xi \rangle$. One can calculate this and analogous correlator $\langle y^2 \xi \rangle$ using the Shapiro-Loginov approximation (2.8), which leads to

$$\frac{d}{dx} \langle E^2 \xi \rangle = 2 \langle Ey\xi \rangle - \lambda \langle E^2 \xi \rangle; \quad (10.23)$$

$$\frac{d}{dx} \langle y^2 \xi \rangle = -2\varepsilon_0 \langle Ey\xi \rangle - 2\varepsilon_1^2 \sigma^2 \langle Ey \rangle - \lambda \langle y^2 \xi \rangle + 2a \langle y\xi \rangle \sin kx.$$

Repeating again the Shapiro-Loginov procedure for the correlator $\langle Ey\xi \rangle$, one gets finally,

$$\frac{d}{dx} \langle Ey\xi \rangle = \langle y^2 \xi \rangle - \varepsilon_0 \langle E^2 \xi \rangle - \varepsilon_1^2 \sigma^2 \langle E^2 \rangle - \lambda \langle Ey\xi \rangle + a \langle E\xi \rangle \sin kx. \quad (10.24)$$

In obtaining (10.23) and (10.24) we used the assumption of symmetric dichotomous noise which allowed as, to replace ξ^2 by σ^2 inside the averages.

Six Equations (10.21)-(10.24) can be reduced to the following two third-order differential equations for $\langle E^2 \rangle$ and $\langle E^2 \xi \rangle$,

$$\frac{d^3}{dx^3} \langle E^2 \rangle + 4\varepsilon_0 \frac{d}{dx} \langle E^2 \rangle + 4\varepsilon_1 \frac{d}{dx} \langle E^2 \xi \rangle + 2\lambda \varepsilon_1 \langle E^2 \xi \rangle =$$

$$4a \langle y \rangle \sin kx + 2A \frac{d}{dx} (\langle E \rangle \sin kx), \tag{10.25}$$

and

$$\frac{d^3}{dx^3} \langle E^2 \xi \rangle + 3\lambda \frac{d^2}{dx^2} \langle E^2 \xi \rangle + \left(4\varepsilon_0 + 3\lambda^2\right) \frac{d}{dx} \langle E^2 \xi \rangle + \lambda \left(4\varepsilon_0 + \lambda^2\right) \langle E^2 \xi \rangle +$$

$$4\varepsilon_1^2 \sigma^2 \frac{d}{dx} \langle E^2 \rangle + 2\lambda \varepsilon_1^2 \sigma^2 \langle E^2 \rangle = 2a\lambda \langle E\xi \rangle \sin kx + 4a \langle y\xi \rangle \sin kx +$$

$$2a \frac{d}{dx} (\langle E\xi \rangle \sin kx) \tag{10.26}$$

where $\langle E \rangle$, $\langle y \rangle$, $\langle E\xi \rangle$, and $\langle y\xi \rangle$ are given in (10.17)-(10.19).

One can proceed now as in the calculation of the second moments, by excluding the correlator $\langle E^2 \xi \rangle$ from Equations (10.25)-(10.26) and by solving the obtained six order differential equation for $\langle E^2 \rangle$, but we do not bring here these cumbersome formulas.

10.3.1 *Driven Maxwell equation*

Let us consider now response to an external field. To this end, we put the zero boundary conditions in Equations (10.25)-(10.26). Moreover, we will not write the cumbersome formulas for the Laplace transform of the right hand side of these equations, restricting ourselves only to the asymptotic solutions at $x \to \infty$ ($p \to 0$). Then, Equations (10.25)-(10.26) are reduced to

$$2\lambda \varepsilon_1^2 \sigma^2 \langle E^2 \rangle_{as} + \lambda \left(\lambda^2 + 4\varepsilon_0\right) \langle E^2 \xi \rangle_{as} = 2a \left[(\lambda \langle E\xi \rangle + 2 \langle y\xi \rangle) \sin kx\right]_{as}, \tag{10.27}$$

$$2\lambda \varepsilon_1 \langle E^2 \xi \rangle_{as} = 4a (\langle y \rangle \sin kx)_{as}$$

which shows [105] that the asymptotic value of the second moment $\langle E^2 \rangle_{as}$, induced by the external field, is a non-monotonic function of σ^2 and λ. Thus, the energy of the output signal, which is proportional to $\langle E^2 \rangle_{as}$, depends non-monotonically both on σ^2 and λ.

10.4 Stochastic resonance in the oscillator with random frequency

In Chapter 7 we analyzed stochastic resonance in overdamped system. Here we consider underdamped equation. The second-order differential equation for driven harmonic oscillation with random frequency (1.13) can be written as two first-order differential equations

$$\frac{dx}{dt} = y, \tag{10.28}$$

$$\frac{dy}{dt} = -2\gamma y - \omega^2 x - \omega^2 \xi x + a \sin \Omega t. \tag{10.29}$$

Equations (10.28) and (10.29), after averaging, take the following form:

$$\frac{d}{dt} \langle x \rangle = \langle y \rangle \tag{10.30}$$

$$\frac{d}{dt} \langle y \rangle = -2\gamma \langle y \rangle - \omega^2 \langle x \rangle - \langle \xi(t) x \rangle + a \sin(\Omega t). \tag{10.31}$$

The new correlator $\langle \xi(t) x \rangle$ has to be found separately. To this end, we use the Shapiro-Loginov procedure [29]). Multiplying Equation (10.28) by ξ, one gets after averaging,

$$\left\langle \xi(t) \frac{dx}{dt} \right\rangle = \langle \xi(t) y \rangle. \tag{10.32}$$

Inserting (10.32) into (2.8) results in

$$\frac{d \langle \xi(t) x \rangle}{dt} = \langle \xi(t) y \rangle - \lambda \langle \xi(t) x \rangle. \tag{10.33}$$

Using the procedure analogous to (10.32)-(10.33) for the correlator $\langle \xi(t) y \rangle$, one gets

$$\frac{d \langle \xi(t) y \rangle}{dt} = \left\langle \xi(t) \frac{dy}{dt} \right\rangle - \lambda \langle \xi(t) y \rangle. \tag{10.34}$$

Multiplying Equation (10.29) by ξ and averaging, one obtains

$$\left\langle \xi \frac{dy}{dt} \right\rangle = -2\gamma \langle \xi y \rangle - \omega^2 \langle \xi x \rangle - \omega^2 \langle \xi^2 x \rangle. \tag{10.35}$$

Equation (10.35) contains the higher order correlator $\langle \xi^2 x \rangle$, and one has to use a decoupling procedure. For dichotomous noise $\xi^2 = \sigma^2$, and Equation (10.35) can be rewritten as

$$\left\langle \xi \frac{dy}{dt} \right\rangle = -2\gamma \langle \xi y \rangle - \omega^2 \langle \xi x \rangle - \omega^2 \sigma^2 \langle x \rangle. \tag{10.36}$$

Inserting (10.36) into (10.34) results in

$$\frac{d \langle \xi(t) y \rangle}{dt} = -2\gamma \langle \xi y \rangle - \omega^2 \langle \xi(t) x \rangle - \omega^2 \sigma^2 \langle x \rangle - \lambda \langle \xi(t) y \rangle . \qquad (10.37)$$

We thus obtain a system of four equations, (10.30), (10.31), (10.33) and (10.37), for four variables, $\langle x \rangle$, $\langle y \rangle$, $\langle \xi x \rangle$, and $\langle \xi y \rangle$.

From these equations one can easily find the fourth-order differential equation for $\langle x \rangle$,

$$\frac{d^4 \langle x \rangle}{dt^4} + 2 (\lambda + 2\gamma) \frac{d^3 \langle x \rangle}{dt^3} + \left(2\omega^2 + \lambda^2 + 6\lambda\gamma + 4\gamma^2 \right) \frac{d^2 \langle x \rangle}{dt^2} +$$

$$\left[2\omega^2 (\lambda + 2\gamma) + 2\lambda\gamma (\lambda + 2\gamma) \right] \frac{d \langle x \rangle}{dt} + \left[\omega^2 \left(\omega^2 + \lambda^2 + 2\lambda\gamma \right) - \omega^2\sigma^2 \right] \langle x \rangle =$$

$$\left(\omega^2 + \lambda^2 + 2\lambda\gamma - \Omega^2 \right) a \sin (\Omega t) + 2 (\lambda + \gamma) a\Omega \cos (\Omega t) . \qquad (10.38)$$

We seek the solution of Equation (10.38) in the form

$$\langle x \rangle = \langle x \rangle_0 + \langle x \rangle_A \qquad (10.39)$$

where the output signal $\langle x \rangle_A$ is induced by an external field $a \sin (\Omega t)$ and $\langle x \rangle_0$ is defined by internal dynamics. For purposes of this discussion, we ignore the possible instability of an underdamped oscillator for adiabatic fluctuations [163].

Let us write the solution $\langle x \rangle_A$ of Equation (10.38) in the form

$$\langle x \rangle_A = A \sin (\Omega t + \phi) \qquad (10.40)$$

Then, one easily finds [14]

$$A = \left[\frac{f_1^2 + f_2^2}{f_3^2 + f_4^2} \right]^{\frac{1}{2}} ; \quad \phi = \tan^{-1} \left(\frac{f_1 f_3 - f_2 f_4}{f_1 f_4 + f_2 f_3} \right) \qquad (10.41)$$

where

$$f_1 = 2 (\lambda + \gamma) \Omega; \quad f_2 = \left(\Omega^2 - \omega^2 - \lambda^2 - 2\lambda\gamma \right) ;$$

$$f_3 = 2 \left(\Omega^2 - \omega^2 \right) \left(\Omega^2 - \omega^2 - \lambda^2 \right) - \omega^2\sigma^2 - \left(6\lambda\gamma + 4\gamma^2 \right) \Omega^2 + 2\lambda\gamma\omega^2;$$

$$\qquad (10.42)$$

$$f_4 = \Omega (\lambda + 2\gamma) \left[\left(\omega^2 - \Omega^2 \right) + \lambda\gamma \right] .$$

In the absence of friction, $\gamma = 0$, Equations (10.41)-(10.42) take the following form

$$A = a \left[\frac{\left[\left(\Omega^2 - \omega^2 - \lambda^2 \right)^2 + 4\lambda^2\Omega^2 \right]}{\left[\left(\Omega^2 - \omega^2 \right) \left(\Omega^2 - \omega^2 - \lambda^2 \right) - \omega^2\sigma^2 \right]^2 + 4\lambda^2\Omega^2 \left(\Omega^2 - \omega^2 \right)^2} \right]^{\frac{1}{2}}$$

$$\qquad (10.43)$$

and

$$\phi = \tan^{-1}\frac{2\Omega\lambda\omega^2\sigma^2}{(\Omega^2 - \omega^2)\left[(\Omega^2 - \omega^2 - \lambda^2)^2 + 4\Omega^2\lambda^2\right] - (\Omega^2 - \omega^2 - \lambda^2)\omega^2\sigma^2}.$$
(10.44)

For small σ^2 Equation (10.43) reduces to Equation (8.6) of [169] found in a different context by perturbation theory.

Before analyzing of Equations (10.43)-(10.44), let us consider the limiting case of white Gaussian noise which corresponds to $\sigma^2 \to \infty$ and $\lambda \to \infty$ with a constant ratio. Then,

$$A = \frac{a}{(\omega^2 - \Omega^2)}; \quad \phi = 0.$$
(10.45)

Amplitude of the output signal A reaches a maximum at

$$\sigma^2 = (\Omega^2 - \omega^2)(\Omega^2 - \omega^2 - \lambda^2)/\omega^2.$$
(10.46)

In Figure 10.1 we show the dependence of the amplitude of a stationary signal on the correlation rate λ for $a = \sigma^2 = \omega = 1$, $\gamma = 0$ and different frequencies Ω of the external field. This graph shows typical SR non-monotonic behavior for $\Omega < \omega$. With a rise in Ω the positions of the maxima are shifted to higher λ, while the heights are non-monotonic functions of Ω.

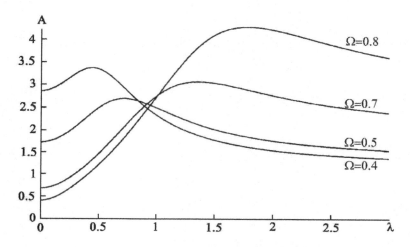

Fig. 10.1 The amplitude of a stationary signal as The amplitude of a stationary signsl as a function of the correlation rate for parameters $\alpha = \sigma^2 = \omega = 1$ and $\gamma = 0$. The curves displayed correspond to different values of the frequency of an external field. [Reprinted figure with permission from Ref. 14. Copyright (2003) by the American Physical Society.]

Analogously, one can find the second moment, $\langle x^2 \rangle$, the correlation function, $\langle x(t_1) x(t_2) \rangle$ and its Fourier expansion, which determines the signal-to-noise ration. The latter was calculated for two white noises for a bistable system [170] and two dichotomous noises [171] for a bilinear system with dichotomous correlations between noises, as well as for a system with quadratic additive noise [172].

Chapter 11

Harmonic Oscillator with Random Damping

11.1 First moment for random damping

11.1.1 Force-free oscillator

Let us first consider the free motion of an oscillator, rewriting for this purpose Equation (1.14) with $m = 1$, and $A = 0$, as

$$\frac{d^2x}{dt^2} + 2\gamma\left[1 + \xi\left(t\right)\right]\frac{dx}{dt} + \omega^2 x = 0 \tag{11.1}$$

or, in the equivalent form,

$$L\{x\} = -2\gamma\xi\frac{dx}{dt}; \quad L\{x\} \equiv \left(\frac{d^2}{dt^2} + 2\gamma\frac{d}{dt} + \omega^2\right)x. \tag{11.2}$$

In order to convert the differential Equation (11.1) into an integro-differential equation we apply, following [165], the operator L^{-1} to the first equation in (11.2), which gives

$$x = -L^{-1}\left\{2\gamma\xi\frac{dx}{dt}\right\}. \tag{11.3}$$

Using the fact that $L\left[L^{-1}\{f\}\right] \equiv f$, one can easily check that the integral operator L^{-1} inverse to the differential operator L defined in (11.2) has the following form

$$L^{-1}\{f\} \equiv \frac{1}{\omega_1}\int_0^t dt_1 \exp[-\gamma(t-t_1)]\sin[\omega_1(t-t_1)]f(t_1); \quad \omega_1 = \sqrt{\omega^2 - \gamma^2}, \tag{11.4}$$

i.e., according to (11.2),

$$x\left(t\right) = -\frac{2\gamma}{\omega_1}\int_0^t dt_1 \exp\left[-\gamma\left(t-t_1\right)\right]\sin\left[\omega_1\left(t-t_1\right)\right]\xi\left(t_1\right)\frac{dx}{dt}\left(t_1\right) \tag{11.5}$$

and

$$\frac{dx}{dt} = \frac{2\gamma}{\omega_1} \int_0^t dt_1 \exp\left[-\gamma\left(t - t_1\right)\right] \xi\left(t_1\right) \frac{dx}{dt}\left(t_1\right) \{\gamma \sin\left[\omega_1\left(t - t_1\right)\right] - \omega_1 \cos\left[\omega_1\left(t - t_1\right)\right]\}. \tag{11.6}$$

On substituting (11.6) into the right hand side of Equation (11.2), one obtains

$$\left(\frac{d^2}{dt^2} + 2\gamma\frac{d}{dt} + \omega^2\right) x\left(t\right) =$$

$$-\frac{4\gamma^2}{\omega_1} \int_0^t dt_1 \exp\left[-\gamma\left(t - t_1\right)\right] \xi\left(t\right) \xi\left(t_1\right) \frac{dx}{dt}\left(t_1\right) \{\gamma \sin\left[\omega_1\left(t - t_1\right)\right] -$$

$$\omega_1 \cos\left[\omega_1\left(t - t_1\right)\right]\}. \tag{11.7}$$

On averaging (11.7) for noise one can use the simplest version of the splitting of averages [165]

$$\left\langle \xi\left(t\right) \xi\left(t_1\right) \frac{dx}{dt}\left(t_1\right) \right\rangle = < \left\langle \xi\left(t\right) \xi\left(t_1\right) \right\rangle \left\langle \frac{dx}{dt}\left(t_1\right) \right\rangle. \tag{11.8}$$

The substitution of (11.8) into the averaging Equation (11.7) shows that for white noise (2.4), one gets

$$\left[\frac{d^2}{dt^2} + 2\gamma\left(1 - 2\gamma D\right)\frac{d}{dt} + \omega^2\right] \langle x \rangle = 0, \tag{11.9}$$

i.e., the presence of white noise in the original Equation (11.1) leads to a decrease of damping if $2\gamma D < 1$. Moreover, if $2\gamma D > 1$, i.e., if the noise is sufficiently strong, the effective damping becomes negative, so that the average value of the coordinate x increases in time, which indicates an instability. On the other hand, for the exponentially correlated noise (2.5) one gets [23]

$$\frac{d^2 \langle x \rangle}{dt^2} + 2\gamma\frac{d \langle x \rangle}{dt} + \frac{4\gamma^2\sigma^2}{\omega_1} \int_0^t \exp\left[-\left(\lambda + \gamma\right)\left(t - t_1\right)\right] \{\gamma \sin\left[\omega_1\left(t - t_1\right)\right] -$$

$$\omega_1 \cos\left[\omega_1\left(t - t_1\right)\right]\}\frac{d \langle x \rangle}{dt}\left(t_1\right) dt_1 + \omega^2 \langle x \rangle = 0. \tag{11.10}$$

Application of the Laplace transform

$$x\left(p\right) = \int_0^\infty \langle x \rangle\left(t\right) \exp\left(-pt\right) dt \tag{11.11}$$

to Equation (11.10) yields

$$\frac{\left(p^2 + \omega^2 + 2\gamma p\right)\left[(p+\lambda)(p+\lambda+2\gamma)+\omega^2\right] - 4\sigma^2 p\gamma^2 (p+\lambda)}{(p+\lambda+2\gamma)(p+\lambda)+\omega^2} x\,(p) =$$

$$\frac{(p+2\gamma)(p+\lambda)(p+\lambda+2\gamma) - 4\sigma^2\gamma^2 (p+\lambda)}{(p+\lambda+2\gamma)(p+\lambda)+\omega^2} x\,(t=0) + \frac{dx}{dt}\,(t=0).$$

$$(11.12)$$

One can check the stability of the solution of Equation (11.1) without performing the inverse Laplace transform in (11.12). In the absence of a driving force and for zero initial condition, $x\,(t=0)=0$, the mean solution $\langle x \rangle$ should relax to zero, which means that the solution of the fourth-order polynomial in p on the left hand side of (11.12) must have no roots with a positive real part. According to the Routh-Hurwitz theorem [173], this condition is obeyed for the fourth order equation $\sum_{i=0}^{i=4} a_i x^i = 0$, if the following relations hold between the coefficients a_i:

$$a_i > 0; \quad a_1 a_4 < a_2 a_3; \quad a_0 a_3^2 < a_1 a_2 a_3 - a_1^2 a_4. \tag{11.13}$$

These stability conditions applied to Equation (11.12) take the following form:

$$\sigma^2 < \min\{2\beta + \alpha + (1+\alpha)^2; \ (1+\alpha^{-1})(\alpha+2\beta); \ \frac{1+\alpha}{2+\alpha}[\alpha+2\beta+2(1+\alpha^2)]\}$$

$$(1+\alpha)^2(\delta-\alpha)(\delta+\alpha+2\alpha^2) < \{2(1+\alpha)[\delta+(1+\alpha)^2 - \sigma^2][(1+\alpha)\delta - \alpha\sigma^2] -$$

$$[(1+\alpha)\delta - \alpha\sigma^2]^2\} \tag{11.14}$$

where

$$\alpha = \frac{\lambda}{2\gamma}; \quad \beta = \frac{\omega^2}{4\gamma^2}; \quad \delta = \alpha + 2\beta. \tag{11.15}$$

The slightly cumbersome inequalities (11.14) define the stability conditions in the form of the relations between three parameters, σ^2, α, and β. In the case of white noise ($\sigma^2 \to \infty$, $\lambda \to \infty$ with $D = \sigma^2/\lambda = \text{const.}$) these inequalities are reduced to the previously obtained condition $2D\gamma < 1$.

11.2 Second moment for random damping

The second order differential Equation (1.14) with $m=1$ can be rewritten as two first order differential equations

$$\frac{dx}{dt} = y; \quad \frac{dy}{dt} = -2\gamma\left[1 + \xi\,(t)\right]y - \omega^2 x + A\sin\left(\Omega t\right) + \eta\,(t). \tag{11.16}$$

The additive noise is taken to be white noise

$$\langle \eta(t)\,\eta(t_1) \rangle = 2\alpha\delta(t - t_1) \tag{11.17}$$

with no correlation with the multiplicative noise.

Multiplying the first of Equations (11.16) by $2x$ and the second one by $2y$, and averaging, one gets

$$\frac{d}{dt}\langle x^2 \rangle = 2\langle xy \rangle, \tag{11.18}$$

$$\frac{d}{dt}\langle y^2 \rangle = -4\gamma\langle y^2 \rangle - 4\gamma\langle \xi y^2 \rangle - 2\omega^2\langle xy \rangle + 2A\langle y \rangle \sin(\Omega t) + 4\alpha.$$

Analogously, multiplying Equations (11.16) by y and x, respectively, summing and averaging the sum leads to

$$\frac{d}{dt}\langle xy \rangle = \langle y^2 \rangle - 2\gamma\langle xy \rangle - 2\gamma\langle \xi xy \rangle - \omega^2\langle x^2 \rangle + A\langle x \rangle \sin(\Omega t). \tag{11.19}$$

In deriving (11.18)-(11.19), we have used Equation (11.17) for correlators containing white noise $\eta(t)$, which gives $\langle y\eta(t) \rangle = 2\alpha$ and $\langle x\eta(t) \rangle = 0$. In addition, Equations (11.18)-(11.19) contain new correlators $\langle \xi y^2 \rangle$ and $\langle \xi xy \rangle$. One can calculate these and the analogous correlator $\langle \xi x^2 \rangle$ using the Shapiro-Loginov procedure (2.8) for $g = x^2$, $g = xy$ and $g = y^2$, respectively. Splitting the higher order correlators into lower order ones involves different approximations. We restrict our attention to the simplest case of white and dichotomous noises where the higher order correlators containing ξ^2 can be split by using $\xi^2 = \sigma^2$ so that, for example, $\langle \xi^2 x^2 \rangle = \sigma^2\langle x^2 \rangle$. Finally one gets

$$\frac{d}{dt}\langle \xi x^2 \rangle = 2\langle \xi xy \rangle - \lambda\langle \xi x^2 \rangle;$$

$$\frac{d}{dt}\langle \xi y^2 \rangle = -4\gamma\langle \xi y^2 \rangle - 4\gamma\sigma^2\langle y^2 \rangle - 2\omega^2\langle \xi xy \rangle + 2A\langle \xi y \rangle \sin(\Omega t) - \lambda\langle \xi y^2 \rangle;$$

$$\frac{d}{dt}\langle \xi xy \rangle = \langle \xi y^2 \rangle - 2\gamma\sigma^2\langle xy \rangle - 2\gamma\langle \xi xy \rangle - \omega^2\langle \xi x^2 \rangle + A\langle \xi x \rangle \sin(\Omega t) - \lambda\langle \xi xy \rangle.$$

$$\tag{11.20}$$

We thus obtain a system of six Equation (11.18)-(11.20) for six variables, $\langle x^2 \rangle$, $\langle y^2 \rangle$, $\langle xy \rangle$, $\langle \xi x^2 \rangle$, $\langle \xi y^2 \rangle$ and $\langle \xi xy \rangle$.

In the simple case of white noise one can use the simplified splitting procedure (2.10) for correlators $\langle \xi y^2 \rangle$, $\langle \xi xy \rangle$, in Equations (11.18)-(11.19) which gives

$$\langle \xi y^2 \rangle \approx -4\gamma D\langle y^2 \rangle; \quad \langle \xi xy \rangle \approx -2\gamma D\langle xy \rangle. \tag{11.21}$$

Substituting (11.21) into (11.18)-(11.19), one obtains

$$\frac{d}{dt}\left\langle x^2\right\rangle = 2\left\langle xy\right\rangle;$$

$$\frac{d}{dt}\left\langle y^2\right\rangle = -4\gamma\left\langle y^2\right\rangle + 16\gamma^2 D\left\langle y^2\right\rangle - 2\omega^2\left\langle xy\right\rangle + 2A\left\langle y\right\rangle\sin\left(\Omega t\right) + 4\alpha;$$

$$\frac{d}{dt}\left\langle xy\right\rangle = \left\langle y^2\right\rangle - 2\gamma\left\langle xy\right\rangle + 4\gamma^2 D\left\langle xy\right\rangle - \omega^2\left\langle x^2\right\rangle + A\left\langle x\right\rangle\sin\left(\Omega t\right). \quad (11.22)$$

Note that for white noise one can find exact expressions for higher moments of the coordinate and velocity of the form $\left\langle x^{m-n}y^n\right\rangle$. The derivative of the latter expression has a form

$$\frac{d}{dt}\left\langle x^{m-n}y^n\right\rangle = (m-n)\left\langle x^{m-n-1}y^n\frac{dx}{dt}\right\rangle + n\left\langle x^{m-n}y^{n-1}\frac{dy}{dt}\right\rangle. \quad (11.23)$$

On substituting (11.16) with $A = 0$ into (11.23), one obtains

$$\frac{d}{dt}\left\langle x^{m-n}y^n\right\rangle = (m-n)\left\langle x^{m-n-1}y^{n+1}\right\rangle - 2n\gamma\left(1 - 2\gamma Dn\right)\left\langle x^{m-n}y^n\right\rangle -$$
$$n\omega^2\left\langle x^{m-n+1}y^{n-1}\right\rangle + nA\sin\left(\Omega t\right)\left\langle x^{m-n}y^{n-1}\right\rangle + 2n\left(n-1\right)\alpha\left\langle x^{m-n}y^{n-2}\right\rangle. \quad (11.24)$$

Equation (11.24) reduces to (11.22) for the special cases $n = 0, m = 2$, $n = 2$, $m = 2$ and $n = 1$, $m = 2$, as it should be.

The correlation functions can be found along the same lines as was done in (11.18)-(11.22) for the second moments by multiplying the first of Equations (11.16) by $x(t_1)$, the second one by $y(t_1)$, and averaging the resulting equations, which gives

$$\frac{d}{dt}\left\langle x(t_1)x(t)\right\rangle = \left\langle x(t_1)y(t)\right\rangle;$$

$$\frac{d}{dt}\left\langle x(t_1)y(t)\right\rangle = -2\gamma\left\langle x(t_1)y(t)\right\rangle - 2\gamma\left\langle \xi(t)x(t_1)y(t)\right\rangle -$$
$$\omega^2\left\langle x(t_1)x(t)\right\rangle + A\left\langle x(t_1)\right\rangle\sin\left(\Omega t\right). \quad (11.25)$$

The new correlators $\left\langle \xi(t)x(t_1)y(t)\right\rangle$ and $\left\langle \xi(t)x(t_1)x(t)\right\rangle$ can be found by using Equation (3.12) leading to

$$\frac{d}{dt}\left\langle \xi(t)x(t_1)x(t)\right\rangle = \left\langle \xi(t)x(t_1)y(t)\right\rangle - \lambda\left\langle \xi(t)x(t_1)x(t)\right\rangle;$$

$$\frac{d}{dt}\left\langle \xi(t)x(t_1)y(t)\right\rangle = -2\gamma\left\langle \xi(t)x(t_1)y(t)\right\rangle - 2\gamma\sigma^2\left\langle x(t_1)y(t)\right\rangle -$$
$$\omega^2\left\langle \xi(t)x(t_1)x(t)\right\rangle - \lambda\left\langle \xi(t)x(t_1)y(t)\right\rangle + A\left\langle \xi(t)x(t_1)\right\rangle\sin\left(\Omega t\right). \quad (11.26)$$

In the following sections, the formulas obtained above will be applied to different special cases.

11.3   Force-free oscillator

11.3.1   *White noise*

The stationary $(d/dt... = 0)$ second moments of the force-free $(A = 0)$ oscillator are obtained from Equations (11.22) which gives

$$\langle xy \rangle_{eq} = 0; \quad \langle y^2 \rangle_{eq} = \omega^2 \langle x^2 \rangle; \quad \langle x^2 \rangle = \frac{\alpha}{2\omega^2 \gamma (1 - 2\gamma D)}. \qquad (11.27)$$

For vanishing multiplicative noise, $D = 0$, Equations (11.27) reduce to their standard form describing Brownian motion. In the presence of this noise, Equations (11.27) show that an (energetic) instability occurs when $2\gamma D > 1$. This result is different from that obtained for the case of the random frequency, where, according to (10.6), the instability condition of the second moment has the form $\omega^2 D > 2\gamma$.

11.3.2   *Dichotomous noise*

For dichotomous noise it is necessary to use the six Equations (11.18)-(11.20), which we will consider in the equilibrium state $(d/dt... = 0)$ and in the absence of an external force $(A = 0)$. Their solution is

$$\langle x^2 \rangle_{eq} = \frac{4\alpha\{(\lambda + 2\gamma)\left[4\omega^2 + \lambda(\lambda + 4\gamma)\right] + 8\lambda\gamma^2\sigma^2\}}{4\gamma(\lambda + 2\gamma)\omega^2\left[4\omega^2 + \lambda(\lambda + 4\gamma)\right] - 8\omega^2\gamma^2\sigma^2\left[2\omega^2 + \lambda(\lambda + 4\gamma)\right]}. \qquad (11.28)$$

In the limit case of white noise $(\sigma^2 \to \infty, \lambda \to \infty$ with $\frac{\sigma^2}{\lambda} = D)$ Equation (11.28) reduces to (11.27) as it should do. For dichotomous noise, the second moment (11.28) is a non-monotonic function of both σ^2 and λ.

11.3.3   *Poisson noise*

For Poisson noise, like for dichotomous noise, one uses Equations (11.18)-(11.20). Using the replacement (2.22), one obtains [174],

$$\langle \xi y \rangle = -\frac{2\gamma\beta\omega_0^2}{1 + \gamma\omega_0}; \qquad \langle \xi x \rangle = 0. \qquad (11.29)$$

The numerical simulations of Equation (11.1) with Poisson shot noise shows [174] the good agreement with the analytical result (11.29).

11.3.4 *Driven oscillator*

All of the preceding was related to the field-free case which corresponds to Equations (11.16) with $A = 0$. From here on, we are interested in the response of the underdamped harmonic oscillator to a periodic field, and, for simplicity, we present only the case of white noise. The more cumbersome expression for dichotomous noise [23] are not given here.

11.4 Second moments

From Equations (11.22) with $\alpha = 0$, one easily finds the third-order differential equation for $\langle x^2 \rangle$,

$$\frac{d^3 \langle x^2 \rangle}{dt^3} + 2\gamma \left(3 - 10\gamma D\right) \frac{d^2 \langle x^2 \rangle}{dt^2} + 2\left[2\omega^2 + 4\gamma^2 \left(1 - 2\gamma D\right)\left(1 - 4\gamma D\right)\right] \frac{d \langle x^2 \rangle}{dt} +$$

$$8\gamma \left(1 - 4\gamma D\right) \omega^2 \langle x^2 \rangle = 4A \left[\langle y \rangle + 2\gamma \left(1 - 4\gamma D\right) \langle x^2 \rangle\right] \sin\left(\Omega t\right) +$$

$$2A \frac{d}{dt} \left[\langle y \rangle \sin\left(\Omega t\right)\right]. \tag{11.30}$$

The first moments, $\langle x \rangle$ and $\langle y \rangle = d \langle x \rangle / dt$, that enter Eq. (11.30) consist of two parts, $\langle x \rangle_0$ and $\langle x \rangle_a$, where the first one is the solution of the field-free Equation (11.9), and the second one is defined by (11.44). In the stable region ($2\gamma D < 1$), the solution of Eq. (11.9) vanishes as $t \to \infty$, and the stationary solution of Equation (11.30) ($d/dt... = 0$) is obtained on substitution $\langle x \rangle = \langle x \rangle_a = a \sin\left(\Omega t + \phi\right)$ and $\langle y \rangle = d \langle x \rangle / dt = a\Omega \cos\left(\Omega t + \phi\right)$. Thus, one gets for the stationary value of the second moment,

$$\langle x^2 \rangle_{st} = \frac{aA}{2\omega^2} \left(\cos\phi - \frac{\Omega \sin\phi}{2\gamma \left(1 - 2\gamma D\right)}\right) \tag{11.31}$$

where a and $\tan\phi$ are defined in the next section ((11.45)-(11.46)). It follows from (11.31) that the second moment $\langle x^2 \rangle_{st}$, is a non-monotonic function of the noise strength D. It turns out that for dichotomous noise the second moment shows a non-monotonic dependence of both σ^2 and λ.

11.5   Correlation functions

From Equations (11.25)-(11.26) one can find the fourth order differential equation for $z \equiv \langle x(t_1) x(t) \rangle$,

$$\frac{d^4 z}{dt^4} + 2(\lambda + 2\gamma) \frac{d^3 z}{dt^3} + \left[2\omega^2 + 4\gamma^2 + 6\lambda\gamma + \lambda^2 - 4\gamma^2 \langle \xi^2 \rangle \right] \frac{d^2 z}{dt^2} +$$

$$\left[(2\omega^2 + 2\lambda\gamma)(\lambda + 2\gamma) - 4\lambda\gamma^2 \langle \xi^2 \rangle \right] \frac{dz}{dt} + \omega^2 \left[\omega^2 + \lambda(\lambda + 2\gamma) \right] z =$$

$$-A \left[\frac{d^2}{dt^2} + 2(\lambda + \gamma)\frac{d}{dt} + (\omega^2 + \lambda^2 + 2\lambda\gamma) \right] \langle x(t_1) \rangle \sin(\Omega t) +$$

$$2A\gamma \left(\frac{d}{dt} + \lambda \right) \langle \xi(t) x(t_1) \rangle \sin(\Omega t). \tag{11.32}$$

In order to avoid the need for the rather long formulas, we restrict our attention to the case of white noise. Then, one gets from Equations (11.26),

$$\langle \xi(t) x(t_1) x(t) \rangle = 0; \quad \langle \xi(t) x(t_1) y(t) \rangle = -2\gamma D \langle x(t_1) y(t) \rangle, \tag{11.33}$$

and,the substitution of these formulae into (11.25) leads to

$$\left[\frac{d^2}{dt^2} + 2\gamma(1 - 4\gamma D)\frac{d}{dt} + \omega^2 \right] \langle x(t_1) x(t) \rangle = Aa \sin(\Omega t_1 + \phi) \sin(\Omega t). \tag{11.34}$$

The form of the solution of Equation (11.34) depends on the type of the initial conditions, which can be either $\langle x(t_1) x(t) \rangle$ at $t = t_1$ or at $t = 0$. For the latter case one assumes that the initial condition $x(t = 0)$ is not random. Then, $\langle x(t_1) x(t = 0) \rangle = x(t = 0) \langle x(t_1) \rangle$ and $\frac{d}{dt} \langle x(t_1) x(t = 0) \rangle = \frac{dx}{dt}(t = 0) \langle x(t_1) \rangle$. Solving the non-homogeneous Equation (11.34) through the use of the Green function, one obtains

$$\langle x(t_1) x(t) \rangle = \exp\left[-\gamma(1 - 4\gamma D)t \right] \{ x(t = 0) \langle x(t_1) \rangle \left[\cos(\omega_2 t) + \right.$$

$$\frac{\gamma(1 - 4\gamma D)}{\omega_2} \sin(\omega_2 t) + \frac{1}{\omega^2} \frac{dx}{dt}(t = 0) \langle x(t_1) \rangle \sin(\omega_2 t)] \} +$$

$$Aa \sin(\Omega t_1 + \phi) \int_0^t \exp\left[-\gamma(1 - 4\gamma D)(t - \theta) \right] \cos\left[\omega_2(t - \theta) \right] \sin(\Omega \theta) \, d\theta \tag{11.35}$$

where $\omega_2 = \sqrt{\omega^2 - \gamma^2 (1 - 4\gamma D)^2}$, and the last integral in (11.35) can be easily expressed in terms of elementary functions.

For white noise, the correlation function (11.35), like the second moment $\langle x^2 \rangle$, is a non-monotonic function of the noise strength D. For dichotomous noise, the correlation function shows a non-monotonic dependence on both σ^2 and λ.

11.6 Stochastic resonance in the oscillator with random damping

Equation (1.14) with an external periodic force $a \sin(\Omega t)$ can be rewritten as two first order differential equations

$$\frac{dx}{dt} = y; \quad \frac{dy}{dt} = -2\gamma y - 2\gamma \xi y - \omega^2 x + a \sin(\Omega t) \tag{11.36}$$

which, after averaging, take the following form

$$\frac{d}{dt} \langle x \rangle = \langle y \rangle; \tag{11.37}$$

$$\frac{d}{dt} \langle y \rangle = -2\gamma \langle y \rangle - 2\gamma \langle \xi y \rangle - \omega^2 \langle x \rangle + a \sin(\Omega t).$$

Equation (11.37) contains a new correlator $\langle \xi y \rangle$ which has to be found separately. To this end, we use the Shapiro-Loginov procedure (2.8) which yields

$$\frac{d}{dt} \langle \xi y \rangle = \left\langle \xi \frac{dy}{dt} \right\rangle - \lambda \langle \xi y \rangle. \tag{11.38}$$

Multiplying the second of Equations (11.36) by ξ, one gets after averaging

$$\left\langle \xi \frac{dy}{dt} \right\rangle = -2\gamma \langle \xi y \rangle - 2\gamma \langle \xi^2 y \rangle - \omega^2 \langle \xi x \rangle. \tag{11.39}$$

Equation (11.39) contains two new correlators, $\langle \xi x \rangle$ and $\langle \xi^2 y \rangle$. The former can be easily found using Equation (2.8), namely

$$\frac{d}{dt} \langle \xi x \rangle = \left\langle \xi \frac{dx}{dt} \right\rangle - \lambda \langle \xi x \rangle = \langle \xi y \rangle - \lambda \langle \xi x \rangle. \tag{11.40}$$

To find the higher-order correlator $\langle \xi^2 y \rangle$ one has to use the splitting procedure, $\langle \xi^2 y \rangle = \langle \xi^2 \rangle \langle y \rangle = \sigma^2 \langle y \rangle$. Note that this procedure becomes exact for the special case of dichotomous noise. In order to keep our calculation exact, we restrict our attention, like the authors of [165], to dichotomous noise, while for the general case of color noise one has to use some approximations. On substituting (11.39) into (11.38) one gets

$$\frac{d}{dt} \langle \xi y \rangle = -2\gamma \langle \xi y \rangle - 2\sigma^2 \gamma \langle y \rangle - \omega^2 \langle \xi x \rangle - \lambda \langle \xi y \rangle. \tag{11.41}$$

We thus obtain a system of four Equations (11.38)-(11.41) for the four variables, $\langle x \rangle$, $\langle y \rangle$, $\langle \xi x \rangle$, and $\langle \xi y \rangle$. From these equations one can easily find the fourth-order differential equation for $\langle x \rangle$

$$\frac{d^4 \langle x \rangle}{dt^4} + 2\left(\lambda + 2\gamma\right) \frac{d^3 \langle x \rangle}{dt^3} + \left[2\omega^2 + \left(\lambda + 2\gamma\right)^2 + 2\lambda\gamma - 4\gamma^2\sigma^2\right] \frac{d^2 \langle x \rangle}{dt^2} +$$

$$2\left[\left(\omega^2 + \gamma\lambda\right)\left(\lambda + 2\gamma\right) - 2\lambda\gamma^2\sigma^2\right] \frac{d \langle x \rangle}{dt} + \omega^2 \left[\omega^2 + \lambda\left(\lambda + 2\gamma\right)\right] \langle x \rangle =$$

$$a\left[\omega^2 - \Omega^2 + \lambda\left(\lambda + 2\gamma\right)\right] \sin\left(\Omega t\right) + 2a\Omega\left(\lambda + \gamma\right) \cos\left(\Omega t\right). \qquad (11.42)$$

We seek the solution of Equation (11.42) in the form

$$\langle x \rangle = \langle x \rangle_0 + \langle x \rangle_A, \qquad (11.43)$$

where the output signal $\langle x \rangle_A$ is induced by an external field, $a \sin\left(\Omega t\right)$, and $\langle x \rangle_0$ is defined by the internal dynamics. Let us write the solution $\langle x \rangle_A$ of the non-homogeneous Equation (11.42) in the form

$$\langle x \rangle_A = A \sin\left(\Omega t + \phi\right). \qquad (11.44)$$

Then, one easily finds that

$$A^2 = \frac{a^2 \left(f_2^2 + \Omega^2\lambda_2^2\right)}{\left(2\Omega^2\gamma\lambda_2 - f_1 f_2 - 4\gamma^2\Omega^2\sigma^2\right)^2 + 4\Omega^2 \left[\lambda_1 \left(\gamma\lambda - f_1\right) - 2\lambda\gamma^2\sigma^2\right]^2} \qquad (11.45)$$

and

$$\tan\phi = \frac{2\Omega\gamma f_2^2 + 4\Omega^3\gamma^2\lambda_2 - 4\Omega\gamma^2\sigma^2 \left(\lambda_1\Omega^2 + \lambda\omega^2 + \lambda^2\lambda_1\right)}{f_1 \left(f_2^2 + \Omega^2\lambda_2^2\right) + 4\Omega^2\gamma^2\sigma^2 \left(f_1 + \lambda^2\right)} \qquad (11.46)$$

where

$$\lambda_1 = \lambda + 2\gamma; \quad \lambda_2 = 2\left(\lambda + \gamma\right); \quad f_1 = \Omega^2 - \omega^2; \quad f_2 = f_1 - \lambda\lambda_1. \qquad (11.47)$$

It follows from (11.45) that the amplitude of the output signal (11.44) shows a non-monotonic dependence on σ^2 and λ (stochastic resonance). The amplitude A reaches a maximum at the following value of the noise strength,

$$\left(\sigma^2\right)_{\max} = \frac{2\gamma \left(\lambda_2\Omega^2 + \lambda_1\lambda^2\right) - \left(\Omega^2 - \omega^2\right) \left(\Omega^2 - \omega^2 - \lambda\lambda_1\right)}{4\gamma^2 \left(\Omega^2 + \lambda^2\right)}. \qquad (11.48)$$

The dependence of the squared ratio A/a of the amplitude of the response signal to that of the external field on the correlation rate λ for $\omega = 2\gamma = 1$ and different frequencies Ω of the external field is shown in Figures 11.1 and 11.2 for two different noise strength $\sigma^2 = 1$ and $\sigma^2 = 5$. These graphs show typical stochastic resonance, with non-monotonic behavior for the frequencies Ω close to the resonance frequency $\Omega = \omega = 1$. The maxima

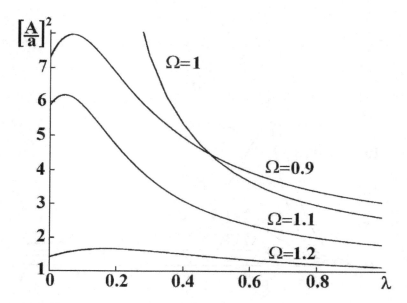

Fig. 11.1   The squared ratio of the amplitude of the response signal to that of an external signal as a function of the correlated rate for noise strength $\sigma^2 = 1$. The curves displayed correspond to different values of the frequency of an external field. [Reprinted figure with permission from Ref. 23. Copyright (2003) by the American Physical Society.]

are more pronounced for larger noise strength. In the limit case of white Gaussian noise, $\sigma^2 \to \infty, \lambda \to \infty$ and $\sigma^2/\lambda = D$, Equation (11.45) takes the following form:

$$A = a \left[\left(\Omega^2 - \omega^2 \right)^2 + 4\gamma^2\Omega^2 \left(1 - 2D\gamma \right)^2 \right]^{-\frac{1}{2}} . \qquad (11.49)$$

The latter result can be also obtained directly from Equation (11.36). Hence, in the presence of white noise one obtains a "dynamic" resonance slightly renormalized by white noise. The amplitude of the output signal A turns out to be a non-monotonic function of the noise strength D for white noise as well, reaching its maximum at $4D = \gamma^{-1}$. The situation becomes more complicated for color noise where the real "stochastic" resonance occurs. For the resonant frequency $\omega = \Omega$ and $2\gamma = 1$, the amplitude of the output signal (11.45) takes the form

$$\frac{A^2}{a^2} = \frac{\lambda^2 \left(\lambda + 1 \right)^2 + \Omega^2 \left(2\lambda + 1 \right)^2}{\Omega^4 \left(2\lambda + 1 - \sigma^2 \right)^2 + \Omega^2\lambda^2 \left(\lambda + 1 - \sigma^2 \right)^2} \qquad (11.50)$$

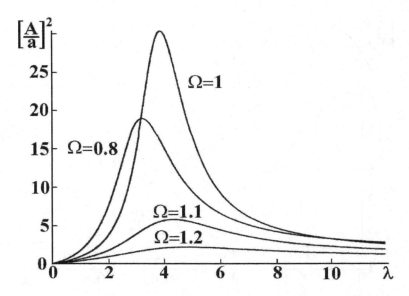

Fig. 11.2   The same as Fig. 11.1 for noise strengths $\sigma^2 = 5$. [Reprinted figure with permission from Ref. 23. Copyright (2003) by the American Physical Society.]

which is non-monotonic function of the frequency of an external field Ω, strength σ^2 and rate λ of noise. For the special case $\sigma^2 = 1$, the amplitude increases indefinitely when $\lambda \to 0$.

11.7   Periodically varying damping

It is instructive to compare the influence of multiplicative noise with the effects found in an oscillator with a periodically changed damping parameter described by the following equation

$$\frac{d^2x}{dt^2} + \gamma\left[1 + b\sin\left(\Omega_1 t\right)\right]\frac{dx}{dt} + \omega^2 x = 0. \tag{11.51}$$

As it was already mentioned in section 7.3, Equation (11.51) with periodic coefficients has the Floquet solution in the form [118]

$$x\left(t\right) = \exp\left(\alpha t\right)\psi\left(t\right) = \exp\left(\alpha t\right)\sum_{n=0}^{\infty}\left[A_n\sin\left(\frac{n\Omega_1 t}{2}\right) + B_n\cos\left(\frac{n\Omega_1 t}{2}\right)\right] \tag{11.52}$$

where the periodic function $\psi\left(t\right)$ is expanded in a Fourier series. As is evident from (11.52), $x\left(t\right)$ vanishes at $t \to \infty$ for $\alpha < 0$, diverges for $\alpha > 0$, and remains a bounded periodic function for $\alpha = 0$. Hence, $\alpha = 0$ defines the stability boundary of the stationary solutions of Equation (11.51). On the substitution of (11.52) with $\alpha = 0$ into (11.51), and comparing the harmonics in front of the sine and cosine terms, one obtains the infinite systems of linear equations for A_n and B_n which have non-zero solutions if the infinite determinant of these equations $\Delta(\alpha = 0)$ vanishes, $\Delta(\alpha = 0) = 0$. One has to truncate this determinant at some n, and afterwards to improve the result by taking into account larger values of n. Leaving only terms with $n = 1$, one obtains the following equations

$$\left(\omega^2 - \frac{\Omega_1^2}{4} + \frac{\Omega_1 \gamma b}{4}\right) A_1 + \frac{\Omega_1 \gamma}{2} B_1 = 0;$$

$$-\frac{\Omega_1 \gamma}{2} A_1 + \left(\omega^2 - \frac{\Omega_1^2}{4} - \frac{\Omega_1 \gamma b}{4}\right) B_1 = 0. \tag{11.53}$$

Equations (11.53) have nontrivial solutions if the determinant of these equations vanishes, which gives

$$b = \sqrt{4 + \frac{(4\omega^2 - \Omega_1^2)^2}{\gamma^2 \Omega_1^2}}. \tag{11.54}$$

The stability boundary (11.54) of the solution $x = 0$ has a V form at $b - \Omega_1$ plane with a stable state located inside this curve. Equation (11.54) defines a necessary condition for the non-zero periodic solution of the periodically varying velocity in the same way as the condition $2\gamma D = 1$ defines it for a random velocity. The difference is that in the former case an external field defines the basic frequency Ω_1 of oscillation, while in the latter case the oscillations occur at the oscillator frequency ω.

Chapter 12

Linear vs Quadratic Noise

In the previous chapters we have seen the difference between multiplicative noises manifested in random mass, damping coefficient and frequency. Contrary to the last two cases, the fluctuations of mass cannot be white since a large negative noise, $\xi(t) << 0$, implies a negative mass of the oscillator. One can bypass the problem of negative mass by replacing $\xi(t)$ by a positive random force $\xi^2(t)$, which corresponds to the fact that the mass of the Brownian particle can only increase due to the adhesion of the molecules of the surrounding medium,

$$\left[1 + \xi^2(t)\right] \frac{d^2x}{dt^2} + 2\gamma \frac{dx}{dt} + \omega^2 x = \eta(t). \qquad (12.1)$$

To consider the cases of random frequency and damping induced by quadratic noise, one replaces Equations (1.13) and (1.14), respectively, by

$$\frac{d^2x}{dt^2} + 2\gamma \frac{dx}{dt} + \omega^2\left[1 + \xi^2(t)\right]x = \eta(t) \qquad (12.2)$$

and

$$\frac{d^2x}{dt^2} + 2\gamma\left[1 + \xi^2(t)\right]\frac{dx}{dt} + \omega^2 x = \eta(t). \qquad (12.3)$$

The quadratic noise $\xi^2(t)$ can be written as follows

$$\xi^2(t) = \sigma^2 + \Delta\xi \qquad (12.4)$$

with $\sigma^2 = AB$ and $\Delta = A - B$. For $\xi = A$, one obtains $\xi^2 = AB + (A - B)A = A^2$, and for $\xi = -B$, $\xi^2 = B^2$

The influence of multiple quadratic noise on the first moment of an overdamped oscillator has been studied by Luczka et al. [175], and we extend their analyses to the calculation of the second moments for different type of noises for underdamped oscillators. It turns out [176] that the stationary second moments for random frequency (Equation (1.13)) and random damping (Equation (1.14)) subject to an additive white noise $\eta(t)$ of strength α, have the following form

$$\langle x^2 \rangle_\omega = \frac{\alpha}{\omega^2 (2\gamma - D\omega^2)}; \qquad \langle x^2 \rangle_\gamma = \frac{\alpha}{2\gamma\omega^2(1 - 4\gamma D)} \qquad (12.5)$$

showing the "energetic" instability when $\langle x^2 \rangle$ becomes negative. In the following sections, we compare the impact of linear and quadratic noise on the stability analysis of the mean-square displacement $\langle x^2 \rangle$.

12.1 Brownian motion

We start from the traditional model of Brownian motion, where the Brownian particle is subject to the systematic damping force $-2\gamma v$ and the linear random force $\eta(t)$ or the quadratic random force $\eta^2(t)$,

$$\frac{dv}{dt} + 2\gamma v = \eta(t) \qquad (12.6)$$

$$\frac{dv}{dt} + 2\gamma v = \eta^2(t). \qquad (12.7)$$

We have seen in Chapter 3, that the mean-square displacement $\langle v^2 \rangle$ for the linear noise is defined by Equation (3.11). Let us now turn to the analysis of Equation (12.7), which can be rewritten, using (12.4), as

$$\frac{dv}{dt} + 2\gamma v = \sigma^2 + \Delta\eta. \qquad (12.8)$$

For a stationary state, averaging Equation (12.8) leads to

$$\langle v \rangle = \frac{\sigma^2}{2\gamma}. \qquad (12.9)$$

Multiplying Equation (12.8) by $2v$ and averaging gives, for a stationary state

$$\langle v^2 \rangle = \frac{\sigma^2}{2\gamma} \langle v \rangle + \frac{\Delta}{2\gamma} \langle \eta v \rangle. \qquad (12.10)$$

Multiplying Equation (12.8) by $\eta(t)$ and averaging yields

$$\langle \eta v \rangle = \frac{\Delta\sigma^2}{\lambda + 2\gamma}. \qquad (12.11)$$

Inserting (12.9) and (12.11) into (12.10) gives

$$\langle v^2 \rangle = \frac{\sigma^2}{2\gamma} \left(\frac{\sigma^2}{2\gamma} + \frac{\Delta^2}{\lambda + 2\gamma} \right). \qquad (12.12)$$

As one can see from Equations (3.11) and (12.12), the stationary second moment $\langle v^2 \rangle$ is positive for both linear and quadratic noise, i.e., the system remains stable.

Thus far, we have analyzed "classical" Brownian motion. To analyze the stochastic oscillator framework, let us consider Brownian motion in a parabolic potential $V(x) = \omega_0^2 x^2 / 2$, which is described, for linear noise by Equation (1.11), and, for the quadratic noise, by the dynamic equation

$$\frac{d^2 x}{dt^2} + 2\gamma \frac{dx}{dt} + \omega_0^2 x = \eta^2(t). \tag{12.13}$$

The stationary second moment for Equation (1.11) has the following form [177]

$$\langle x^2 \rangle = \frac{D}{2\gamma \omega_0^2} \frac{\lambda(\lambda + 2\gamma)}{\lambda^2 + 2\gamma\lambda + \omega_0^2}. \tag{12.14}$$

Analogously, one obtains the stationary second moment for Equation (12.13),

$$\langle x^2 \rangle = \frac{D\Delta}{2\gamma \omega_0^2} \frac{\lambda(\lambda + 2\gamma\Delta)}{\lambda^2 + 2\gamma\lambda + \omega_0^2}, \tag{12.15}$$

which is quite similar to (12.14), and both of them are positively definite.

12.2 Harmonic oscillator with random frequency

Equations (12.2)-(12.4) can be rewritten as two first-order differential equations

$$\frac{dx}{dt} = y \tag{12.16}$$

$$\frac{dy}{dt} = -2\gamma y - \omega^2 \left(1 + \sigma^2\right) x - \omega^2 \Delta \xi x + \eta(t). \tag{12.17}$$

Multiplying (12.16) by $2x$ and (12.17) by $2y$ and averaging leads to

$$\frac{d}{dt} \langle x^2 \rangle = 2 \langle xy \rangle \tag{12.18}$$

$$\frac{d}{dt} \langle y^2 \rangle = -4\gamma \langle y^2 \rangle - 2\omega^2 \left(1 + \sigma^2\right) \langle xy \rangle - 2\omega^2 \Delta \langle \xi xy \rangle + 4D. \tag{12.19}$$

Equation (12.19) contains a new correlator $\langle \xi xy \rangle$, which can be evaluated using the Shapiro-Loginov equation (2.8). Multiplying (12.16) by y and (12.17) by x, summing and averaging the sum gives

$$\frac{d}{dt} \langle xy \rangle = \langle y^2 \rangle - 2\gamma \langle xy \rangle - \omega^2 \left(1 + \sigma^2\right) \langle x^2 \rangle - \omega^2 \Delta \langle \xi x^2 \rangle. \tag{12.20}$$

Equation (2.8) take the following forms for $g = x^2$, y^2 and xy,

$$\frac{d}{dt} \left\langle \xi x^2 \right\rangle = 2 \left\langle \xi xy \right\rangle - \lambda \left\langle \xi x^2 \right\rangle \tag{12.21}$$

$$\frac{d}{dt} \left\langle \xi y^2 \right\rangle >= 2 \left\langle \xi y \frac{dy}{dt} \right\rangle - \lambda \left\langle \xi y^2 \right\rangle =$$
$$-4\gamma \left\langle \xi y^2 \right\rangle - 2\omega^2 \left(1 + \sigma^2\right) \left\langle \xi xy \right\rangle - 2\omega^2 \Delta\sigma^2 \left\langle xy \right\rangle - \lambda \left\langle \xi y^2 \right\rangle; \tag{12.22}$$

$$\frac{d}{dt} \left\langle \xi xy \right\rangle = \left\langle \xi y^2 \right\rangle - 2\gamma \left\langle \xi xy \right\rangle - \omega^2 \left(1 + \sigma^2\right) \left\langle \xi x^2 \right\rangle - \omega^2 \Delta\sigma^2 \left\langle x^2 \right\rangle - \lambda \left\langle \xi xy \right\rangle. \tag{12.23}$$

It was assumed in Equation (12.22) that there is no correlation between additive and multiplicative noise, $\langle \xi\eta \rangle = 0$.

For the equilibrium state $(d/dt.. = 0)$, Equation (12.18) yields $\langle xy \rangle = 0$ and Equation (12.21) yields, $\left\langle \xi x^2 \right\rangle = (2/\lambda) \left\langle \xi xy \right\rangle$ Inserting the latter formulas into Equations (12.19), (12.20), (12.22) and (12.23), one obtains,

$$-4\gamma \left\langle y^2 \right\rangle - 2\omega^2 \Delta \left\langle \xi xy \right\rangle + 4D = 0 \tag{12.24}$$

$$\left\langle y^2 \right\rangle - \omega^2 \left(1 + \sigma^2\right) \left\langle x^2 \right\rangle - \frac{2\omega^2\Delta}{\lambda} \left\langle \xi xy \right\rangle = 0 \tag{12.25}$$

$$[4\gamma + \lambda] \left\langle \xi y^2 \right\rangle = -2\omega^2 \left(1 + \sigma^2\right) \left\langle \xi xy \right\rangle \tag{12.26}$$

$$\left\langle \xi y^2 \right\rangle - 2\gamma \left\langle \xi xy \right\rangle - \frac{2\omega^2}{\lambda} \left(1 + \sigma^2\right) \left\langle \xi xy \right\rangle - \omega^2\Delta\sigma^2 \left\langle x^2 \right\rangle - \lambda \left\langle \xi xy \right\rangle = 0. \tag{12.27}$$

Inserting $\left\langle \xi y^2 \right\rangle$ from Equation (12.27) into (12.26) gives

$$\left\{ 2\omega^2(1+\sigma^2) + (4\gamma+\lambda) \left[\lambda + 2\gamma + \frac{2\omega^2}{\lambda}(1+\sigma^2) \right] \right\} \langle \xi xy \rangle = -\omega^2\Delta\sigma^2(4\gamma+\lambda)\langle x^2 \rangle. \tag{12.28}$$

Inserting $\left\langle y^2 \right\rangle$ from (12.25) into (12.24) leads to

$$-4\gamma \left(1 + \sigma^2\right) \omega^2 \left\langle x^2 \right\rangle - \left[4\gamma \frac{2\omega^2\Delta}{\lambda} + 2\omega^2\Delta \right] \left\langle \xi xy \right\rangle + 4D = 0. \tag{12.29}$$

Finally, excluding $\langle \xi xy \rangle$ from (12.28) and (12.29) results in

$$\left\langle x^2 \right\rangle = 4D \left\{ 4\gamma\omega^2 \left(1 + \sigma^2\right) - \frac{2\omega^4\Delta^2\sigma^2 \left(4\gamma + \lambda\right)^2}{(\lambda + 2\gamma) \left[4\omega^2 \left(1 + \sigma^2\right) + (4\gamma + \lambda) \lambda\right]\}} \right\}^{-1} \tag{12.30}$$

which (for $\Delta = 1$ and $1 + \sigma^2 \to 1$) reduces to Equation (10.25) in [176],

$$\langle x^2 \rangle = D \left\{ 4\gamma\omega^2 - \frac{2\omega^4\sigma^2 (4\gamma + \lambda)^2}{(\lambda + 2\gamma)[4\omega^2 + (4\gamma + \lambda)\lambda]} \right\}^{-1}. \tag{12.31}$$

According to Equations (12.30) and (12.31), the stability condition (positivity of $\langle x^2 \rangle$) for quadratic noise has the form

$$\frac{2\omega^2\Delta^2\sigma^2 (4\gamma + \lambda)^2}{2\gamma (1 + \sigma^2)(2\gamma + \lambda)[4\omega^2 (1 + \sigma^2) + \lambda (4\gamma + \lambda)]} < 1 \tag{12.32}$$

and for linear noise

$$\frac{2\omega^2\Delta^2\sigma^2 (4\gamma + \lambda)^2}{2\gamma (2\gamma + \lambda)[4\omega^2 + \lambda (4\gamma + \lambda)]} < 1. \tag{12.33}$$

Comparison of (12.32) and (12.33) shows, that for small strength σ of quadratic noise, like in the case of linear noise, an oscillator becomes unstable when the inequality (12.33) is not obeyed. However, by increasing the strength σ of quadratic noise, one can satisfy (12.32), thus stabilizing the oscillator with the help of quadratic noise (noise-induced stability).

12.3 Harmonic oscillator with random damping

We briefly outline the calculations which are similar to those described in the previous section. Equations (12.3)-(12.4) can be rewritten as

$$\frac{dx}{dt} = y \tag{12.34}$$

$$\frac{dy}{dt} = -2\gamma (1 + \sigma^2) y - \omega^2 x - 2\gamma\Delta\xi y + \eta (t). \tag{12.35}$$

Multiplying (12.34) by $2x$ and (12.35) by $2y$ and averaging,

$$\frac{d}{dt} \langle x^2 \rangle = 2 \langle xy \rangle \tag{12.36}$$

$$\frac{d}{dt} \langle y^2 \rangle = -4\gamma (1 + \sigma^2) \langle y^2 \rangle - 2\omega^2 \langle xy \rangle - 4\gamma\Delta \langle \xi y^2 \rangle + 4D. \tag{12.37}$$

Multiplying (12.34) by y and (12.35) by x, summing and averaging the sum,

$$\frac{d}{dt} \langle xy \rangle = \langle y^2 \rangle - 2\gamma (1 + \sigma^2) \langle xy \rangle - \omega^2 \langle x^2 \rangle - 2\gamma\Delta \langle \xi xy \rangle \tag{12.38}$$

Equations (12.37)-(12.38) contain new correlators $\langle \xi x^2 \rangle$, $\langle \xi y^2 \rangle$ and $\langle \xi xy \rangle$. One can calculate these correlators by multiplying Equations (12.34), (12.35) and (12.38) by $2\xi x$, $2\xi y$ and ξ, respectively, using the Shapiro-Loginov equation (2.8) for $g = x^2$ or y^2, or xy, and averaging,

$$\frac{d}{dt} \langle \xi x^2 \rangle = 2 \langle \xi xy \rangle - \lambda \langle \xi x^2 \rangle \qquad (12.39)$$

$$\frac{d}{dt} \langle \xi y^2 \rangle = -4\gamma \left(1 + \sigma^2\right) \langle \xi y^2 \rangle - 4\gamma \Delta \sigma^2 \langle y^2 \rangle - 2\omega^2 \langle \xi xy \rangle - \lambda \langle \xi y^2 \rangle \qquad (12.40)$$

$$\frac{d}{dt} \langle \xi xy \rangle = \langle \xi y^2 \rangle - \left[\lambda + 2\gamma \left(1 + \sigma^2\right)\right] \langle \xi xy \rangle - \omega^2 \langle \xi x^2 \rangle - 2\gamma \Delta \sigma^2 \langle xy \rangle . \qquad (12.41)$$

We thus obtain a system of six Equations (12.36)-(12.41) for six variables, $\langle x^2 \rangle$, $\langle y^2 \rangle$, $\langle xy \rangle$, $\langle \xi x^2 \rangle$, $\langle \xi y^2 \rangle$ and $\langle \xi xy \rangle$.

For the equilibrium state $(d/dt.. = 0)$, one gets a similar result to that obtained in the previous section,

$$\langle x^2 \rangle = \{D\{2\omega^2 + \lambda[\lambda + 2\gamma(1 + \sigma^2)]\}[\lambda + 4\gamma(1 + \sigma^2)] + 2\omega^2\lambda + 8\lambda\gamma^2\Delta^2\sigma^2\} \times$$

$$\{4\gamma\omega^2(1 + \sigma^2)2\omega^2 + \lambda[\lambda + 2\gamma(1 + \sigma^2)][\lambda + 4\gamma(1 + \sigma^2)] -$$

$$16\omega^2\gamma^2\sigma^2\Delta^2 2\omega^2 + \lambda[\lambda + 2\gamma(1 + \sigma^2)]\}^{-1}. \qquad (12.42)$$

In the limiting case of linear noise, $1 + \sigma^2 \to 1$ and $\Delta = 1$, the latter equation reduces to

$$\langle x^2 \rangle = \frac{D\left\{\left[2\omega^2 + \lambda\left[\lambda + 2\gamma\right]\right]\left(\lambda + 4\gamma\right) + 2\omega^2\lambda + 8\lambda\gamma^2\Delta^2\sigma^2\right\}}{4\gamma\omega^2\left[2\omega^2 + \lambda\left(\lambda + 2\gamma\right)\left(\lambda + 4\gamma\right)\right] - 16\gamma^2\sigma^2\omega^2\left[2\omega^2 + \lambda\left(\lambda + 2\gamma\right)\right]}. \qquad (12.43)$$

The stability condition for linear noise is

$$2\omega^2 + \lambda\left(\lambda + 2\gamma\right)\left(\lambda + 4\gamma\right) > 4\gamma\sigma^2\left[2\omega^2 + \lambda\left(\lambda + 2\gamma\right)\right] \qquad (12.44)$$

and for quadratic noise

$$2\omega^2 + \lambda\left[\lambda + 2\gamma\left(1 + \sigma^2\right)\right]\left[\lambda + 4\gamma\left(1 + \sigma^2\right)\right] > 4\gamma\sigma^2\left[\frac{2\omega^2 + \lambda^2}{1 + \sigma^2} + 2\gamma\lambda\right]. \qquad (12.45)$$

Therefore, as in the previously considered case of random frequency, an increase in the strength of quadratic noise makes a system more stable compared to linear noise.

12.4 Harmonic oscillator with random mass

12.4.1 *Linear noise of small strength ($\sigma^2 < 1$)*

One has to introduce quadratic noise for the case of a random mass because large negative noise leads to negative mass of the Brownian particle. Another possibility exists for small noise strength σ. Multiplying Equation (1.15) by $1 - \xi(t)$, yields, for symmetric dichotomous noise $\xi(t) = \pm\sigma$,

$$\left(1 - \sigma^2\right)\frac{d^2x}{dt^2} + 2\gamma\frac{dx}{dt} + \omega^2 x = \xi(t)\left(2\gamma\frac{dx}{dt} + \omega^2\right)x + \eta(t)\left[1 - \xi(t)\right].$$
(12.46)

Since the oscillator mass must be positive, the condition $\sigma^2 < 1$, must be satisfied in studies of Equation (12.46). Therefore, we approximate $1 - \sigma^2$ by unity, concluding that the small dichotomous fluctuations of mass are equivalent to simultaneous fluctuations of the frequency and the damping coefficient.

To calculate the second moment $\langle x^2 \rangle$, it is convenient to rewrite the second-order differential equation (12.46) as an equivalent system of two first-order differential equations,

$$\frac{dx}{dt} = y; \qquad \frac{dy}{dt} + 2\gamma\frac{dx}{dt} + \omega^2 x - \xi(t)\left(2\gamma\frac{d}{dt} + \omega^2\right)x = \eta(t). \quad (12.47)$$

We assume in Equation (12.47) that there is no correlation between additive and multiplicative noise, $\langle \eta(t)\xi(t) \rangle = 0$

Multiplying the first of Equation (12.47) by $2x$ and the second by $2y$ yields

$$\frac{d}{dt}x^2 = 2xy; \qquad \frac{d}{dt}y^2 + 4\gamma y^2 + 2\omega^2 xy - 4\gamma\xi y^2 - 2\omega^2\xi xy = 4D. \quad (12.48)$$

Averaging Equation (12.48) by using (2.8) yields

$$\frac{d}{dt}\langle x^2 \rangle = 2\langle xy \rangle; \qquad \left(\frac{d}{dt} + 4\gamma\right)\langle y^2 \rangle + \omega^2\frac{d\langle x^2 \rangle}{dt} - 4\gamma\langle\xi y^2\rangle - 2\omega^2\langle\xi xy\rangle = 4D.$$
(12.49)

Analogously, multiplying Equations (12.47) by y and x, respectively, and summing leads to

$$\frac{d}{dt}xy = y^2 - \xi x\frac{dy}{dt} - 2\gamma xy - \omega^2 x^2 + 2\eta x =$$

$$y^2 - \left(\frac{d}{dt} + \lambda\right)\xi xy + \xi y^2 - \xi\eta x - 2\gamma xy - \omega^2 x^2 + 2x\eta. \quad (12.50)$$

We restrict our analysis to stationary moments $((d/dt) \ldots = 0)$, which according to (12.49) and 12.50), leads to

$$\langle xy \rangle = 0; \qquad \lambda \langle \xi y^2 \rangle + 2\gamma \langle y^2 \rangle = 4D; \tag{12.51}$$
$$\langle y^2 \rangle - \lambda \langle \xi xy \rangle + \langle \xi y^2 \rangle - \omega^2 \langle x^2 \rangle = 0.$$

Equations (12.51) contain new correlators $\langle \xi x^2 \rangle$, $\langle \xi y^2 \rangle$ and $\langle \xi xy \rangle$. One can calculate these correlators by multiplying Equations (12.47) and (12.50) by $2\xi x$, $2\xi y$ and ξ, respectively, and averaging,

$$\lambda \langle \xi x^2 \rangle = 2 \langle \xi xy \rangle \tag{12.52}$$

$$(\lambda + 4\gamma) \langle \xi y^2 \rangle + \sigma^2 \frac{d}{dt} \langle y^2 \rangle + 2\omega^2 \langle \xi xy \rangle = 0 \tag{12.53}$$

$$(\lambda + 2\gamma) \langle \xi xy \rangle - \langle \xi y^2 \rangle + \sigma^2 \langle y^2 \rangle + \omega^2 \langle \xi x^2 \rangle = 0. \tag{12.54}$$

We thus obtain six equations, (12.51)-(12.54), for the six variables $\langle x^2 \rangle$, $\langle y^2 \rangle$, $\langle xy \rangle$, $\langle \xi x^2 \rangle$, $\langle \xi y^2 \rangle$, and $\langle \xi xy \rangle$. We will not write down the cumbersome dynamic equations for the second moments, which can easily be obtained from this system of differential equations, but shall restrict our attention to the non-correlated additive and multiplicative noises for small fluctuations of mass $(\sigma^2 < 1)$

$$\langle x^2 \rangle = \frac{4D}{\omega^2} \frac{(2 + \sigma^2) C - \Delta^2 \sigma^2 (1 + \lambda A/\omega^2)}{(4\gamma C - \lambda \Delta^2 \sigma^2)} \tag{12.55}$$

where

$$C = \frac{AB}{2\omega^2} + 2 + \sigma^2 + \frac{A}{\lambda}; \quad A = 2(\lambda + 2\gamma) + (\lambda + \Delta) \sigma^2; \quad B = \lambda + 4\gamma + \lambda \sigma^2. \tag{12.56}$$

For linear noise, $1 + \sigma^2$ and Δ have to be replaced by unity, which results in

$$\langle x^2 \rangle = \frac{4D}{\omega^2} \frac{2C - \sigma^2 (1 + \lambda A/\omega^2)}{(4\gamma C - \lambda \sigma^2)} \tag{12.57}$$

with

$$C = \frac{AB}{2\omega^2} + 2 + \frac{A}{\lambda}; \quad A = 2(\lambda + 2\gamma) + (\lambda + 1) \sigma^2; \quad B = \lambda + 4\gamma + \lambda \sigma^2. \tag{12.58}$$

Similar to the cases of random frequency and random damping, the transition from linear (Equation (12.55)) to quadratic (Equation (12.57)) noise leads to an increase of stability.

We now proceed to the analysis of Equation (12.1), which according to Equation (12.4), can be rewritten as

$$\left[1 + \sigma^2 + \Delta\xi(t)\right]\frac{d^2x}{dt^2} + 2\gamma\frac{dx}{dt} + \omega^2 x = \eta(t), \tag{12.59}$$

or

$$\left[1 + \sigma^2\right]\frac{d^2x}{dt^2} + 2\gamma\frac{dx}{dt} + \omega^2 x = \eta(t) - \Delta\xi(t)\frac{d^2x}{dt^2}. \tag{12.60}$$

Let us first analyze Equation (12.60) within the framework of the linear-response theory.

12.4.2 *Linear-response theory*

According to linear-response theory, the output to the input $\eta(t) - \Delta\xi(t)\left(d^2x/dt^2\right)$ yields

$$x(t) = \frac{1}{\omega_1\left(1 + \sigma^2\right)}\int_0^t du\left[\exp\left(c_1(t-u)\right) - \exp\left(c_2(t-u)\right)\right] \times$$

$$\left[\eta(u) - \Delta\xi(u)\frac{d^2x(u)}{dt^2}\right] \tag{12.61}$$

where

$$\omega_1 = \sqrt{\left(\gamma^2 - \omega^2\right)/\left(1 + \sigma^2\right)}; \qquad c_{1,2} = \left(-\gamma \pm \omega_1\right)/\left(1 + \sigma^2\right). \tag{12.62}$$

Differentiating Equation (12.61) twice with respect to t,

$$\frac{d^2x}{dt^2} = \frac{1}{\omega_1\left(1 + \Delta\sigma^2\right)}\int_0^t du\left[c_1^2\exp\left(c_1(t-u)\right) - c_2^2\exp\left(c_2(t-u)\right)\right] \times$$

$$\left[\eta(u) - \Delta\xi(u)\frac{d^2x(u)}{dt^2}\right]. \tag{12.63}$$

Inserting (12.63) into (12.60) and averaging, leads to

$$\left[1 + \sigma^2\right]\frac{d^2\langle x\rangle}{dt^2} + 2\gamma\frac{d\langle x\rangle}{dt} + \omega^2\langle x\rangle = -\frac{\Delta}{\omega_1\left(1 + \sigma^2\right)}\int_0^t du\{[c_1^2\exp\left(c_1(t-u)\right) -$$

$$c_2^2\exp\left(c_2(t-u)\right)] - \left[\langle\xi(t)\eta(u)\rangle - \Delta\left\langle\xi(t)\xi(u)\frac{d^2x(u)}{dt^2}\right\rangle\right]\}. \tag{12.64}$$

Using the Shapiro-Loginov procedure [29] to split the correlations

$$\left\langle\xi(t)\xi(u)\frac{d^2x(u)}{dt^2}\right\rangle = \langle\xi(t)\xi(u)\rangle\frac{d^2\langle x(u)\rangle}{dt^2}, \tag{12.65}$$

which is exact for the exponentially correlated noise, one obtains for white noise, $< \xi(t) \xi(u) > = D\delta(t-u)$

$$\left[1 + \sigma^2 + \frac{D\Delta\left(c_1^2 - c_2^2\right)}{\omega_1\left(1 + \sigma^2\right)}\right] \frac{d^2\langle x\rangle}{dt^2} + 2\gamma\frac{d\langle x\rangle}{dt} + \omega^2\langle x\rangle = \frac{R\Delta\left(c_1^2 - c_2^2\right)t}{\omega_1\left(1 + \sigma^2\right)}$$

(12.66)

where we assume a delta-correlation between additive noise $\eta(t)$ and multiplicative noise $\xi(t)$

$$\langle \xi(t)\eta(u)\rangle = R\delta(t-u). \tag{12.67}$$

12.4.3   *Quadratic noise*

As in the previous cases, replace Equation (12.59) by two first-order differential equations

$$\frac{dx}{dt} = y; \qquad \frac{dy}{dt} == -\left(1 + \sigma^2\right)\frac{dy}{dt} - \Delta\xi\frac{dy}{dt} - 2\gamma y - \omega^2 x + \eta(t) \quad (12.68)$$

Multiplying the first equation in (12.68) by $2x$ and the second by $2y$ gives, after averaging and using Equation (2.8),

$$\frac{d}{dt}\left\langle x^2\right\rangle = \langle xy\rangle; \tag{12.69}$$

$$\frac{d}{dt}\left\langle y^2\right\rangle = -\left(1 + \sigma^2\right)\frac{d}{dt}\left\langle y^2\right\rangle - \Delta\left(\frac{d}{dt} + \lambda\right)\left\langle \xi y^2\right\rangle - 4\gamma\left\langle y^2\right\rangle - 2\omega^2\langle xy\rangle + 4D.$$

Multiplying Equations (12.68) by y and x, respectively, summing and averaging these equations,

$$\frac{d}{dt}\langle xy\rangle = \left(2 + \sigma^2\right)\left\langle y^2\right\rangle - \Delta\left(\frac{d}{dt} + \lambda\right)\langle \xi xy\rangle + \Delta\left\langle \xi y^2\right\rangle - \omega^2\left\langle x^2\right\rangle. \quad (12.70)$$

For the steady-state $(d/dt... = 0)$, Equations (12.69) and (12.70) reduce to the following form,

$$\langle xy\rangle = 0; \qquad \Delta\lambda\left\langle \xi y^2\right\rangle + 4\gamma\left\langle y^2\right\rangle = 4D; \tag{12.71}$$

$$\left(2 + \sigma^2\right)\left\langle y^2\right\rangle - \Delta\lambda\langle \xi xy\rangle + \Delta\left\langle \xi y^2\right\rangle - \omega^2\left\langle x^2\right\rangle = 0.$$

Multiplying Equations (12.68) and (12.70) by $2x\xi$, $2y\xi$, and ξ, respectively, and averaging, one obtains for the steady-state,

$$\lambda\left\langle \xi x^2\right\rangle = 2\langle \xi xy\rangle; \qquad \left[4\gamma + \lambda\left(2 + \sigma^2\right)\right]\left\langle \xi y^2\right\rangle = -2\omega^2\langle \xi xy\rangle; \quad (12.72)$$

$$(\lambda + 2\gamma)\langle \xi xy\rangle = \left(2 + \sigma^2\right)\left\langle \xi y^2\right\rangle + \Delta\sigma^2\left\langle y^2\right\rangle - \omega^2\left\langle \xi x^2\right\rangle.$$

From the six equations, (12.71)-(12.72), one obtains

$$\left\langle x^2\right\rangle = \frac{2D}{\omega^2}\frac{\left(2 + \sigma^2\right)A - \Delta^2\sigma^2(\lambda + B)}{4\gamma A - \Delta\lambda\sigma^2 B} \tag{12.73}$$

where

$$A = \lambda + 2\gamma + \left(2 + \sigma^2\right) B + \frac{2\omega^2}{\lambda}; \qquad B = \frac{2\omega^2}{4\gamma + \lambda\left(2 + \sigma^2\right)}. \qquad (12.74)$$

For linear noise $2 + \sigma^2 \to 2$ and $\Delta = 1$,

$$\langle x^2 \rangle = \frac{2D}{\omega^2} \frac{2A_1 - \sigma^2\left(\lambda + B_1\right)}{4\gamma A_1 - \lambda\sigma^2 B_1} \qquad (12.75)$$

with

$$A_1 = \lambda + 2\gamma + 2B_1 + \frac{2\omega^2}{\lambda}; \qquad B_1 = \frac{\omega^2}{2\gamma + \lambda}. \qquad (12.76)$$

Comparing Equations (12.73)-(12.76) shows that $A > A_1$ and $B < B_1$, i.e., as in the previous cases, the replacement of linear noise by quadratic noise makes the system more stable.

For white noise, Equation (12.75) reduces to

$$\langle x^2 \rangle = \frac{D}{2\gamma\omega^2}. \qquad (12.77)$$

This result agrees with the well-known result for "free" Brownian motion meaning $\omega^2 = 0$. For a free Brownian particle one obtains $\langle x^2 \rangle \to \infty$, as expected. The fact that the stability is unaffected by the mass fluctuation is due to the fact that the multiplicative random force appears in Equation (12.59) in front of a higher derivative.

Chapter 13

Nonlinear Oscillator with Multiplicative Noise

13.1 Double-well potential (noise induced reentrant transition)

In Chapter 7 we considered an overdamped noisy oscillator in a double-well potential. In this chapter the same problem is discussed for an underdamped oscillator. In the presence of damping, the equation of motion has the form

$$\frac{d^2x}{dt^2} + \gamma\frac{dx}{dt} - ax + bx^3 + \sqrt{D}\xi(t)x = 0. \tag{13.1}$$

It turns out [178], that for a noisy system there is some region of positive a where a reentrant transitions occur as a function of the noise strength, showing noisy oscillations for both weak and strong noise.

On passing to dimensionless variables $\tau = \gamma t$ and $y = \sqrt{b}x/\gamma$, one gets,

$$\frac{d^2y}{d\tau^2} + \frac{dy}{d\tau} - \alpha y + y^3 + \sqrt{\Delta}\xi(t)y = 0 \tag{13.2}$$

with dimensionless coefficients

$$\alpha = \frac{a}{\gamma^2}, \quad \Delta = \frac{D}{\gamma^4}. \tag{13.3}$$

Stability at the origin, $x = 0$, is defined by the linearized version of (13.2)

$$\frac{d^2y}{d\tau^2} + \frac{dy}{d\tau} - \alpha y + \sqrt{\Delta}\xi\left(\gamma^{-1}\tau\right)y = 0. \tag{13.4}$$

The deviations from the trivial solution $y(\tau) = \frac{dy}{d\tau}(\tau) = 0$, which are described by Equation (13.4), will lead to a new nontrivial solution when the Lyapunov index of this equation will change the sign. The Lyapunov index

Λ is defined as the exponential divergence rate of neighboring trajectories [179], i.e., as

$$\Lambda = \lim_{t\to\infty} \lim_{\epsilon\to 0} \frac{1}{t} \int_0^t d\tau \frac{1}{\varepsilon} \ln \frac{y(\tau+\varepsilon)}{y(\tau)}. \tag{13.5}$$

It is convenient, [180] to take the limit $\varepsilon \to 0$ first. Then, after substituting in (13.5) the expansion $y(\tau+\varepsilon) = y(\tau) + \varepsilon\frac{dy}{d\tau} + ...$, one gets

$$\Lambda = \lim_{t\to\infty} \frac{1}{t} \int_0^t d\tau \; z(\tau) = \langle z \rangle \equiv \int z P_{as}(z)\, dz \tag{13.6}$$

where the new variable $z = \frac{dy/d\tau}{y}$ has been introduced. In order to find the asymptotic ($\tau \to \infty$) probability distribution function $P_{as}(z)$ for the variable z, one can, using (13.4), find the Langevin equation for z,

$$\frac{dz}{d\tau} = \alpha - z - z^2 - \sqrt{\Delta}\xi(\tau). \tag{13.7}$$

So far we did not make any assumption about the type of noise in the original Equation (13.1). However, for different types of noise one gets different forms of the Fokker-Planck equation corresponded to the Langevin Equation (13.7).

For a Gaussian white noise $\xi(x)$,

$$\langle \xi(t) \rangle = 0, \quad \langle \xi(x)\xi(x_1) \rangle = 2\delta(x - x_1), \tag{13.8}$$

the Stratonovich form of the Fokker-Planck equation for $P(z)$ which corresponds to the Langevin Equation (13.7) with noise of the form (13.8) has the following form [34]

$$\frac{\partial P}{\partial \tau} = -\frac{\partial}{\partial z}\left[(\alpha - z - z^2)P\right] + \Delta\frac{\partial^2 P}{\partial z^2} = 0. \tag{13.9}$$

The integrable asymptotic ($\tau \to \infty$) solution of this equation is found by the method of variation of constants, which gives

$$P_{as}(z) = N \int_{-\infty}^z \exp\left\{\frac{1}{\Delta}[\Psi(z) - \Psi(y)]\right\} \tag{13.10}$$

where $\Psi(y) = \alpha y - y^2/2 - y^3/3$.

On substituting (13.10) into (13.6), one obtains [178],

$$\Lambda = \frac{1}{2}\left\{\frac{\int_0^\infty du\sqrt{u}\exp\left[(\alpha+\frac{1}{4})u/\Delta - u^4/12\Delta\right]}{\int_0^\infty \frac{du}{\sqrt{u}}\exp\left[(\alpha+\frac{1}{4})u/\Delta - u^4/12\Delta\right]} - 1\right\}. \tag{13.11}$$

For different values of the parameters α and Δ defined in (13.3), the Lyapunov index Λ can be either positive, describing thereby the stability of the origin, $x = 0$, or negative which represents an instability.

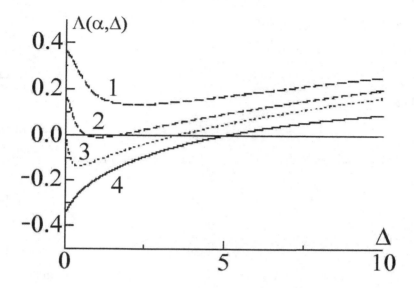

Fig. 13.1   Lyapunov exponent of a linear damped oscillator with parametric white noise as a function of the noiseamplitude for different values of the control parameter α(1: $\alpha = 0.5$, 2: $\alpha = 0.2$, 3: $\alpha = 0.0$, 4: $\alpha = -0.2$). [Reprinted from Ref. 179 with permission from EDP Sciences.]

Analysis of the function $\Lambda\left(\alpha, \Delta\right)$ shown in Figure 13.1 shows that [178]

a) For $\alpha = -0.2$, λ changes sign once (noise-induced instability).
b) For $\alpha = 0.2$, λ changes sign twice (noise induced reentrant transition)
c) For $\alpha = 0.5$, λ is positive for all Δ (no transition).

Hence, noise can suppress oscillations by stabilizing a fixed point which was unstable in the absence of noise. The stability analysis of the linearized form (13.4) of the original Equation (13.1) has to be performed by the use of the Lyapunov exponent which is the proper indicator of the transition in the nonlinear system. Such an analysis cannot be performed by studies of the finite-order moments of the linearized system. Indeed, the second-order moments of Equation (13.4) with white noise are always unstable for positive α which contradicts the results obtained above.

Note that, in addition to above considered white noise, the asymptotic distribution function $P_{as}\left(z\right)$ corresponding to the Langevin Equation (13.7) was found in section 6.1.3 and 6.1.4 for the more general case of dichotomous and Poisson noises. For dichotomous noise (2.6) with $\xi\left(t\right) = \pm\sigma$ and the flipping rate λ, $P_{as}(z)$ is given as

$$P_{as} = N \ (z - z_1)^{-1-\lambda/(1-\sigma^2)} \ |z - z_1 + \alpha - \sigma|^{-1-\lambda/(\sigma^2+1)} \times$$

$$|z - z_1 + \alpha + \sigma|^{-1-\lambda/(\sigma^2-1)} \tag{13.12}$$

where

$$z_1 = \frac{1 - \sqrt{1 + 4\alpha^2}}{2}. \tag{13.13}$$

For Poisson noise (2.19)-(2.20), $P_{as}(z)$ has the following form

$$P_{as} = N \ (z - z_1 + z_2) |z - z_2|^{-1-1/wz_2} \tag{13.14}$$

where

$$z_2 = \lambda w + 1. \tag{13.15}$$

The Lyapunov index Λ is obtained after substitution (13.12) and (13.14) into (13.6), and carrying out the numerical integration.

13.2   Duffing oscillator

The general form of power law nonlinearity in the underdamped oscillator equation of the form

$$\frac{d^2x}{dt^2} + \left[\omega^2 + \xi(t)\right] x + bx^{2n-1} = 0 \tag{13.16}$$

has been analyzed [181]. The linear equation which corresponds to $n = 1$ was considered in Chapter 10. Although some results can be obtained for the general case [181], we restrict ourselves to the special case of Equation (13.16) with $n = 2$ which is called the Duffing oscillator , described by the following equation

$$\frac{d^2x}{dt^2} + \gamma \frac{dx}{dt} + \left[\omega^2 + \xi(t)\right] x + bx^3 = 0. \tag{13.17}$$

Note that the equation of motion (13.17) of the Duffing oscillator is distinguished from the Equation (13.1) for the motion in a double-well potential (which is sometimes called an "inverted Duffing oscillator") by the sign in front of the linear term, $\omega^2 x$. This leads to an important difference between these two equations, namely that the deterministic and undamped Duffing oscillator ($\gamma = \xi = 0$ in (13.17)) has only one stable point, $x = 0$.

The latter equation allows an exact solution since, according to (13.17), the energy

$$E = \frac{1}{2}\left(\frac{dx}{dt}\right)^2 + \frac{\omega^2 x^2}{2} + \frac{bx^4}{4} \tag{13.18}$$

is conserved. A solution of (13.17) in terms of Jacobi elliptic functions is obtained in the energy-angle variables. Turning now to the general Equation (13.17), one gets [178] the following result. Starting from a small initial condition (small energy), the amplitude of the oscillator grows exponentially with time as long as $E \ll \omega^4$ or $x \ll \omega$. For $E \gg \omega^4$, the nonlinear term becomes important, and for white noise the amplitude grows as the square root of time. The crossover from exponential to algebraic growth appears at $E \sim \omega^4$ or $x \sim \omega$, where the linear and nonlinear terms are of the same order. It is remarkable that the scaling indices defining the asymptotic time dependences $x \approx t^\alpha$ and $E \approx t^\beta$ are different for white and colored noise $\xi(t)$ in Equation (13.17) in such a way that for white noise $\alpha = 1/2$ and $\beta = 2$ while for colored noise these indices are two times smaller, $\alpha = 1/4$ and $\beta = 1$.

The analysis of stability for Equation (13.17) can be performed analogously to that of (13.1)-(13.11) in the previous chapter. Although in this case the origin remains the fixed point for all parameters α and Δ defined in (13.3), the Lyapunov index Λ which is equal to [178]

$$\Lambda = \frac{1}{2}\left\{ \frac{\int_0^\infty du\sqrt{u}\exp\left[(1 - \alpha^2/4)\,2u/\Delta - u^3/6\Delta\right]}{\int_0^\infty \frac{du}{\sqrt{u}}\exp\left[(1 - \alpha^2/4)\,2u/\Delta - u^3/6\Delta\right]} - \alpha \right\} \tag{13.19}$$

defines the transition from the absorbing state at origin characterized by delta function form of stationary distribution to an oscillatory asymptotic state with a nontrivial form of the stationary distribution function representing a dynamic balance between energy dissipation and noise-induced energy injection. The transitions between these two states are defined by the sets of parameters α and Δ in (13.19) which make the Lyapunov index Λ equal zero.

13.3   Van der Pol oscillator

The van der Pol oscillator is described by the following equation [182]

$$\frac{d^2x}{dt^2} + \gamma\left(x^2 - 1\right)\frac{dx}{dt} + \omega_0^2 x = 0. \tag{13.20}$$

The nonlinear damping term in (13.20) describes the energy exchange with a surrounding medium in such a way that energy is dissipated when $|x| > 1$, and fed-in when $|x| < 1$. This intrinsic periodicity is a mechanism responsible for the fact that all but one initial conditions will lead to closed stable trajectories in the phase space x, $\frac{dx}{dt}$ (limit cycles) The only other asymptotic state is a point $x = \frac{dx}{dt} = 0$ which will be achieved only if one starts from this point. All other initial conditions will lead to limit cycles.

Consider first an additive constant force acting on van der Pol oscillator [183]

$$\frac{d^2x}{dt^2} + \left(x^2 - 1\right)\frac{dx}{dt} + x = A \qquad (13.21)$$

where by using the appropriate units for x and t, we put $\omega_0^2 = \gamma = 1$.

The bias force A in (13.21) perturbs the intrinsic periodicity of the system and introduces a new fixed point $x = A$, $\frac{dx}{dt} = 0$, which turns out to be stable for $|A| > 1$. For $|A| < 1$, an asymptotic solution, as in (13.20), has the form of a limit cycle, i.e., the bifurcation point is $x = A_c = 1$, $\frac{dx}{dt} = 0$.

Let us generalize Equation (13.21) by introducing the white noise $\xi(t)$,

$$\frac{d^2x}{dt^2} + \left(x^2 - 1\right)\frac{dx}{dt} + x = A + \sqrt{D}\xi(t). \qquad (13.22)$$

The transition from (13.21) to (13.22) means that the control parameter A is now fluctuating, and the stochastic bifurcation process between the dissipative and oscillating mechanisms is different from the deterministic one. It turns out that [183]

1. For both deterministic and noisy systems the bifurcation point A_c, which divides two types of solutions has a marginal stability, which means that it takes a relaxation time $\tau \sim |A - A_c|^{-1}$ to approach this point (critical slowing-down similar to that in equilibrium systems near the critical points).

2. The location of bifurcation points A_c slightly depends on the noise intensity D, $A_c = f(D)$ in such a way that limit cycles exist up to A larger than $A_c = 1$ for $D = 0$.

3. The latter phenomenon is connected with the noise-induced periodicity caused by white noise (without an external periodic signal!) which is different from the internal periodicity of limit cycles of deterministic systems.

Stochastic noise can be added to the deterministic system (13.20) not as additive noise, as in (13.22), but as multiplicative noise

$$\frac{d^2x}{dt^2} + \gamma \left(x^2 - 1 \right) \frac{dx}{dt} + \omega_0^2 x + \sqrt{2D} \xi \left(t \right) x = 0. \tag{13.23}$$

An interesting phenomenon has been found from the analysis of this Equation [184]. Since the deterministic parameter ω_0^2 determines the period of a simple harmonic oscillator and strongly influences the period of limit cycles, its randomness leads to a new phenomena, namely the period of oscillations is found to be a decreasing function of noise intensity. This phenomenon of noise-induced speeding-up is in contrast to previously mentioned critical slowing-down.

Let us consider now the van der Pol oscillator subjected to an external periodic field $A \sin (\Omega t)$,

$$\frac{d^2x}{dt^2} + \gamma \left(x^2 - 1 \right) \frac{dx}{dt} + \omega_0^2 x = A \sin (\Omega t). \tag{13.24}$$

The solution of Equation (13.24) is, in general, a sum of two periodic solutions corresponding to the natural frequency of the oscillator ω_0 and the frequency of the driving force Ω. When Ω approaches ω_0, the phenomenon of frequency locking occurs, namely, the solution of (13.24) with the frequency ω_0 will disappear at some finite value of $|\Omega - \omega_0|$ (and not at $\Omega = \omega_0$), and the remaining solution will be that with the frequency Ω. Performing the first approximation of the Krylov-Bogolyubov averaging procedure [185], one can show [182] that for $A^2/\gamma^2\Omega^2 > 32/27$, the frequency locking solution is stable for the following relation between the amplitude A and frequency Ω of the external field,

$$\mid \Omega - \omega_0 \mid < \frac{\gamma}{4} \sqrt{\left(\frac{2A^2}{\gamma^2\Omega^2} - 1 \right)}. \tag{13.25}$$

For $A^2/\gamma^2\Omega^2 < 1$, the frequency locking solution is stable for all values of Ω up to a threshold value of Ω_c found in [182]. An approximate analysis of the frequency locked solution has also been performed in the presence of additive white noise $\xi (t)$ to Equation (13.24) [182].

Chapter 14

Harmonic Oscillator with Random Mass

Comparing Equations (1.15) and (1.23) shows that the solution of Equation (1.15) can be obtained from the solution of Equation (1.23) by the replacement $1 + \sigma^2$ and Δ by unity. Therefore, we will consider the stochastic Equation (1.23), which can be rewritten as an equivalent system of two first-order differential equations,

$$\frac{dx}{dt} = v; \qquad \frac{dv}{dt} = -\sigma^2 \frac{dv}{dt} - \Delta\xi \frac{dv}{dt} - 2\gamma v + \eta(t) + A\cos(\omega t). \qquad (14.1)$$

Averaging Equations (14.1) and using (2.8) gives

$$\frac{d\langle x \rangle}{dt} = \langle v \rangle; \qquad (1 + \sigma^2)\frac{d\langle v \rangle}{dt} + \Delta\left(\frac{d}{dt} + \lambda\right)\langle \xi v \rangle + 2\gamma\langle v \rangle = A\cos(\omega t).$$
$$(14.2)$$

For the stationary state ($d/dt... = 0$), one gets from Equations (14.2), $\langle v \rangle = 0$; $\langle \xi v \rangle = (A/\Delta\lambda)\cos(\omega t)$ and from Equation (2.8) with $g = v$, $\langle \xi\eta(dv/dt)\rangle = \lambda\langle \xi\eta v \rangle$. Multiplying Equation (14.1) by η and averaging, one gets for stationary states,

$$\langle \eta v \rangle = \frac{D}{2\gamma + (1 + \sigma^2)\lambda} - \frac{\Delta\lambda}{2\gamma + (1 + \sigma^2)\lambda}\langle \xi\eta v \rangle = \frac{D}{2\gamma + (1 + \sigma^2)\lambda};$$
$$(14.3)$$

$$\langle \eta x \rangle = \frac{D}{\lambda[2\gamma + (1 + \sigma^2)\lambda]}.$$

In the last equations we have used the simplest splitting of the correlators,

$$\langle \xi\eta v \rangle = \langle \xi\eta \rangle \langle v \rangle = 0. \qquad (14.4)$$

Additional relations between averaged values can be obtained by multiplying the first equation in (14.1) by $2x$ and the second by $2v$, and averaging the resulting equations by using Equation (2.8),

$$\frac{d}{dt}\langle x^2 \rangle = 2 \langle xv \rangle ; \tag{14.5}$$

$$\frac{d}{dt}\langle v^2 \rangle = -\sigma^2 \frac{d}{dt}\langle v^2 \rangle - \Delta\left(\frac{d}{dt}+\lambda\right)\langle \xi v^2 \rangle - 4\gamma\langle v^2 \rangle + 2\langle v\eta \rangle .$$

Multiplying Equations (14.1) by v and x, respectively, and summing these equations gives

$$\frac{d}{dt}(xv) = (1+\sigma^2)v^2 - \sigma^2 \frac{d}{dt}(xv) - \Delta\xi x\frac{dv}{dt} - 2\gamma xv + x\eta(t) + xA\cos(\omega t). \tag{14.6}$$

Averaging Equation (14.6) results in

$$\frac{d}{dt}\langle xv \rangle = \left(1+\sigma^2\right)\langle v^2 \rangle - \Delta\left(\frac{d}{dt}+\lambda\right)\langle \xi xv \rangle + \Delta\langle \xi v^2 \rangle -$$
$$\left(\sigma^2\frac{d}{dt}+2\gamma\right)\langle xv \rangle + \frac{D}{\lambda\left[2\gamma+(1+\sigma^2)\lambda\right]} + \langle x \rangle A\cos\left(\omega t\right). \tag{14.7}$$

For the stationary state, Equations (14.5) and (14.7) can be reduced to the following form,

$$\langle xv \rangle = 0; \qquad \Delta\lambda\langle \xi v^2 \rangle + 4\gamma\langle v^2 \rangle = \frac{2D}{\left[2\gamma+(1+\sigma^2)\lambda\right]}; \tag{14.8}$$

$$\left(1+\sigma^2\right)\langle v^2 \rangle - \Delta\lambda\langle \xi xv \rangle + \Delta\langle \xi v^2 \rangle + \frac{D}{\lambda\left[2\gamma+(1+\sigma^2)\lambda\right]} + \langle x \rangle A\cos\left(\omega t\right)=0.$$

Multiplying Equations (14.1) and (14.6) by $2x\xi$, $2\xi v$ and ξ, respectively, and averaging, one obtains for the stationary state,

$$\lambda\langle \xi x^2 \rangle = 2\langle \xi xv \rangle ;$$

$$\left[4\gamma+\lambda\left(1+\sigma^2\right)\right]\langle \xi v^2 \rangle + \Delta\sigma^2\langle v^2 \rangle - 2\langle \xi v \rangle A\cos\left(\omega t\right) = 2\langle v\xi\eta \rangle = 0;$$

$$\left[2\gamma+\left(1+\sigma^2\right)\lambda\right]\langle \xi xv \rangle - \left(1+\sigma^2\right)\langle \xi v^2 \rangle - \Delta\sigma^2\langle v^2 \rangle$$

$$\Delta\left\langle \xi^2\frac{d}{dt}(xv)\right\rangle + \langle \xi x \rangle A\cos\left(\omega t\right) = \langle x\xi\eta \rangle = D_1\langle x \rangle . \tag{14.9}$$

Equations (14.8) and (14.9) yield

$$\langle v^2 \rangle\left[4\gamma - \frac{\lambda\Delta^2\sigma^2}{4\gamma+\lambda\left(1+\sigma^2\right)}\right] = \frac{2D}{2\gamma+(1+\sigma^2)\lambda} - \frac{A^2}{4\gamma+\lambda\left(1+\sigma^2\right)} \tag{14.10}$$

where, for the long time behavior, the additional average has been performed, $\langle\cos^2(\omega t)\rangle = 1/2$.

For linear noise, Equation (14.10) becomes

$$\frac{2D}{\lambda(\lambda + 2\gamma)} - \frac{A^2}{\lambda[4\gamma + \lambda]} = \langle v^2 \rangle \left[\frac{4\gamma}{\lambda} - \frac{\sigma^2}{4\gamma + \lambda} \right], \tag{14.11}$$

which, for $A = 0$ and $\sigma^2 = 0$, reduces to

$$\langle v^2 \rangle = \frac{\lambda}{4\gamma} \frac{2D}{\lambda(\lambda + 2\gamma)} = \frac{D}{2\gamma(\lambda + 2\gamma)} \tag{14.12}$$

and coincides with Equation (3.11) for the usual Brownian motion, as required.

In summary, for the Brownian motion with adhesion in the absence of an external field, $A = 0$, the second moment $\langle v^2 \rangle$ is equal to

$$\langle v^2 \rangle \left[4\gamma - \frac{\lambda \Delta^2 \sigma^2}{4\gamma + \lambda(1 + \sigma^2)} \right] = \frac{2D}{2\gamma + (1 + \sigma^2)\lambda} \tag{14.13}$$

i.e., $\langle v^2 \rangle$ becomes negative for large strength σ^2 of the mass fluctuations showing instability of a system, i.e., a system cannot reach the stationary state. In the presence of an external field, the large amplitude of this field might be the alternative source of the instability. These results are attributable to the violating of the energetic balance (fluctuation-dissipation theorem) for the usual Brownian motion. Note that a system becomes stable if both the strength of the mass fluctuations and the amplitude of an external field are sufficiently large, which restores the energetic balance.

We expect that the model of a Brownian particle with fluctuating mass, as well as the model previously considered of an oscillator with random mass [22], will find many applications in modern science.

14.1 Basic equations

For generality we consider trichotomous noise, when for the stationary states, the probabilities P of values $\pm a$ and 0 are

$$P(-a) = P(a) = q; \qquad P(0) = 1 - 2q. \tag{14.14}$$

The limit case of dichotomous noise corresponds to $q = 1/2$. The supplementary conditions to the Ornstein-Uhlenbeck correlations are

$$\xi^3 = a^2 \xi; \qquad \xi^2 = 2qa^2. \tag{14.15}$$

Equation (1.15) with $m = 1$ can be rewritten as two first-order differential equations

$$\frac{dx}{dt} = y; \qquad \frac{dy}{dt} = -\xi \frac{dy}{dt} - \gamma y - \omega^2 x + A \sin(\Omega t) \tag{14.16}$$

which after averaging take the following form

$$\frac{d \langle x \rangle}{dt} = \langle y \rangle \,; \qquad \frac{d \langle y \rangle}{dt} = - \left(\frac{d}{dt} + \lambda \right) \langle \xi y \rangle - \gamma \langle y \rangle - \omega^2 \langle x \rangle + A \sin (\Omega t)$$

(14.17)

where the Shapiro-Loginov formula for splitting the correlation [29] (with $n = 1$),

$$\left\langle \xi (t) \frac{d^n g}{dt^n} \right\rangle = \left(\frac{d}{dt} + \lambda \right)^n \langle \xi g \rangle \,,$$

(14.18)

which yields for exponentially correlated noise, has been used.

Additional relation between averaged values can be obtained by multiplying the first of Equation (14.16) by $2x$ and the second by $2y$ yields

$$\frac{d}{dt} x^2 = 2xy; \qquad \frac{d}{dt} y^2 + \xi \frac{dy^2}{dt} + 2\gamma y^2 + 2\omega^2 xy = 2y A \sin (\Omega t) \,. \qquad (14.19)$$

Averaging Equations (14.19) by using (14.18) yields

$$\frac{d}{dt} \langle x^2 \rangle = 2 \langle xy \rangle \,;$$

(14.20)

$$\left(\frac{d}{dt} + 2\gamma \right) \langle y^2 \rangle + \left(\frac{d}{dt} + \lambda \right) \langle \xi y^2 \rangle + 2\omega^2 \langle xy \rangle = 2 \langle y \rangle A \sin (\Omega t) \,.$$

Analogously, multiplying Equations (14.16) by y and x, respectively, and summing leads to

$$\frac{d}{dt} xy = y^2 - \xi \left[\frac{d}{dt} (xy) - y^2 \right] - \gamma xy - \omega^2 x^2 + 2x A \sin (\Omega t)$$

$$= y^2 - \left(\frac{d}{dt} + \lambda \right) \xi xy + \xi y^2 - \gamma xy - \omega^2 x^2 + 2x A \sin (\Omega t) \quad (14.21)$$

which yields after averaging and using (14.18),

$$\frac{d}{dt} \langle xy \rangle = \langle y^2 \rangle - \left(\frac{d}{dt} + \lambda \right) \langle \xi xy \rangle + \langle \xi y^2 \rangle - \gamma \langle xy \rangle - \omega^2 \langle x^2 \rangle + 2 \langle x \rangle A \sin (\Omega t) \,.$$

(14.22)

Additional equations for the correlators can be obtained by multiplying Equations (14.16) and (14.21) by $2\xi x, 2\xi y$ and ξ, respectively, and averaging,

$$\left(\frac{d}{dt} + \lambda \right) \langle \xi x^2 \rangle = 2 \langle \xi xy \rangle$$

(14.23)

$$\left(\frac{d}{dt} + \lambda + 2\gamma \right) \langle \xi y^2 \rangle + 2\omega^2 \langle \xi xy \rangle = 2 \langle \xi y \rangle A \sin (\Omega t)$$

(14.24)

$$\left(\frac{d}{dt} + \lambda + \gamma\right)\langle\xi xy\rangle = \langle\xi y^2\rangle - \omega^2\langle\xi x^2\rangle + \omega^2\sigma^2\langle x^2\rangle + 2\langle\xi y\rangle A\sin(\Omega t)$$

$$(14.25)$$

which, in the long time limit, $t \to \infty$ are reduced to

$$\lambda\langle\xi x^2\rangle = 2\langle\xi xy\rangle\,; \qquad (\lambda + 2\gamma)\langle\xi y^2\rangle + 2\omega^2\langle\xi xy\rangle = 2\langle\xi y\rangle A\sin(\Omega t)\,;$$

$$(\lambda + \gamma)\langle\xi xy\rangle = \langle\xi y^2\rangle - \omega^2\langle\xi x^2\rangle + \omega^2\sigma^2\langle x^2\rangle + 2\langle\xi y\rangle A\sin(\Omega t)\,.$$

$$(14.26)$$

In the limit $t \to \infty$, Equations'(14.22), (14.23), (14.24) and (14.25) take the following form

$$\lambda\langle\xi y\rangle + \omega^2\langle x\rangle = A\sin(\Omega t) \tag{14.27}$$

$$2\gamma\langle y^2\rangle + \lambda\langle\xi y^2\rangle = 0 \tag{14.28}$$

$$\langle y^2\rangle - \lambda\langle\xi xy\rangle + \langle\xi y^2\rangle - \gamma\langle xy\rangle - \omega^2\langle x^2\rangle + 2\langle x\rangle A\sin(\Omega t) = 0. \tag{14.29}$$

Multiplying Equations (14.16) by $\xi(t)$ and afterwards averaging results in

$$\left(\frac{d}{dt} + \lambda\right)\langle\xi x\rangle = \langle\xi y\rangle\,;$$

$$\left(\frac{d}{dt} + \lambda\right)\langle\xi y\rangle = -\left\langle\xi^2\left[-\xi\frac{dy}{dt} - \gamma y - \omega^2 x + A\sin(\Omega t)\right]\right\rangle - \gamma\langle\xi y\rangle - \omega^2\langle\xi x\rangle =$$

$$a^2\left(\frac{d}{dt} + \lambda\right)\langle\xi y\rangle + \gamma\langle\xi^2 y\rangle + \omega^2\langle\xi^2 x\rangle - 2qAa^2\sin(\Omega t) - \gamma\langle\xi y\rangle - \omega^2\langle\xi x\rangle\,.$$

$$(14.30)$$

In the second equation in (14.30) we used the Shapiro-Loginov formula (14.18) for $\langle\xi\rangle = 0$, which for our case takes the forms (2.13)-(2.14).

Finally, for the stationary states, Equation (14.30) take the form

$$\langle\xi^2 y\rangle - \lambda\langle\xi^2 x\rangle + 2q\lambda a^2\langle x\rangle = 0 \tag{14.31}$$

and

$$-a^2\lambda\langle\xi y\rangle - (\gamma + \lambda)\langle\xi^2 y\rangle -$$

$$\omega^2\langle\xi^2 x\rangle + 2q\lambda a^2\langle y\rangle + 2qAa^2\sin(\Omega t) = 0 \tag{14.32}$$

By this means we obtained ten equations, (14.26), (14.27), (14.31), (14.29), (14.28), (14.32) for ten correlators $\langle x\rangle$, $\langle x^2\rangle$, $\langle y^2\rangle$, $\langle\xi x\rangle$, $\langle\xi y\rangle$, $\langle\xi xy\rangle$, $\langle\xi x^2\rangle$, $\langle\xi y^2\rangle$, $\langle\xi^2 x\rangle$ and $\langle\xi^2 y\rangle$.

14.2   First moment

In order to find $\langle x \rangle$ from the equations, obtained in the previous section, let us first exclude $\langle \xi y \rangle$ from these equations, which gives

$$\lambda^2 \langle \xi x \rangle + \omega^2 \langle x \rangle = A \sin{(\Omega t)} \tag{14.33}$$

$$\lambda \left[\lambda \left(1 - a^2 \right) + \gamma \right] + \omega^2 \langle \xi x \rangle = \gamma \langle \xi^2 y \rangle + \omega^2 \langle \xi^2 x \rangle - 2qAa^2 \sin{(\Omega t)} \tag{14.34}$$

$$2\gamma \langle y^2 \rangle + \lambda \langle \xi^2 y \rangle = A \sin{(\Omega t)} \tag{14.35}$$

$$\lambda \langle \xi x^2 \rangle = 2 \langle \xi x y \rangle; \qquad (\lambda + 2\gamma) \langle \xi y^2 \rangle + 2\omega^2 \langle \xi x y \rangle = 2\lambda \langle \xi x \rangle A \sin{(\Omega t)}. \tag{14.36}$$

Inserting $\langle \xi^2 y \rangle$ from Equation (14.34) into (14.35) and (14.36) results in

$$\frac{\lambda \left\{ \lambda \left[\lambda \left(1 - a^2 \right) + \gamma \right] + \omega^2 \right\}}{\gamma} \langle \xi x \rangle - \frac{\omega^2 \lambda}{\gamma} \langle \xi^2 x \rangle + 2\gamma \langle y^2 \rangle +$$
$$\frac{2q\lambda Aa^2}{\gamma} \sin{(\Omega t)} = A \sin{(\Omega t)} \tag{14.37}$$

$$- \left\{ \frac{(2\gamma + \lambda) \left\{ \lambda \left[\lambda \left(1 - a^2 \right) + \gamma \right] + \omega^2 \right\}}{\gamma} \right\} \langle \xi x \rangle - \frac{\omega^2 (2\gamma + \lambda)}{\gamma} \langle \xi^2 x \rangle -$$
$$\frac{2qa^2 (2\gamma + \lambda)}{\gamma} A \sin{(\Omega t)} + \omega^2 \lambda \langle \xi x^2 \rangle = 2\lambda \langle \xi x \rangle A \sin{(\Omega t)}. \tag{14.38}$$

Substituting $\langle \xi x \rangle$ from (14.33) into (14.37) and (14.38) leads to

$$\frac{\omega^2 \left\{ \lambda \left[\lambda \left(1 - a^2 \right) + \gamma \right] + \omega^2 \right\}}{\gamma \lambda} \langle x \rangle + \frac{\left\{ \lambda \left[\lambda \left(1 - a^2 \right) + \gamma \right] + \omega^2 \right\}}{\gamma \lambda} A \sin{(\Omega t)} -$$
$$\frac{\omega^2 \lambda}{\gamma} \langle \xi^2 x \rangle + 2\gamma \langle y^2 \rangle + \frac{2q\lambda a^2}{\gamma} A \sin{(\Omega t)} = A \sin{(\Omega t)} \tag{14.39}$$

$$\frac{\omega^2 (2\gamma + \lambda) \left\{ \lambda \left[\lambda \left(1 - a^2 \right) + \gamma \right] + \omega^2 \right\}}{\gamma \lambda^2} \langle x \rangle -$$

$$\frac{(2\gamma + \lambda) \left\{ \lambda \left[\lambda \left(1 - a^2 \right) + \gamma \right] + \omega^2 \right\}}{\gamma \lambda^2} A \sin{(\Omega t)} - \frac{\omega^2 (2\gamma + \lambda)}{\gamma} \langle \xi^2 x \rangle -$$
$$\frac{2qa^2 (2\gamma + \lambda)}{\gamma} A \sin{(\Omega t)} + \omega^2 \lambda \langle \xi x^2 \rangle = 2\lambda \langle \xi x \rangle A \sin{(\Omega t)}. \tag{14.40}$$

Finally one obtains

$$\langle x \rangle = \frac{\lambda^3 \alpha}{\gamma \omega^2 \left\{ B\omega^2 \lambda - \gamma \lambda \left(\omega^2 + \lambda \gamma \right) \left[B \left(\lambda + \gamma \right) - a^2 \lambda^2 \right] - 2q\lambda^4 a^2 \right\}} +$$

$$\left\{ \frac{2q\lambda a^2}{\omega^2} + \frac{B}{\lambda^2} - \frac{\left(\omega^2 + \lambda \gamma \right) \left[B \left(\gamma + \lambda \right) - a^2 \lambda^2 \right]}{\omega^2 \lambda^3 \gamma} \right\} A \sin \left(\Omega t \right) \quad (14.41)$$

where

$$B = \frac{\lambda \left\{ \left[\lambda \left(1 - a^2 \right) - \gamma \right] + \omega^2 \right\}}{\gamma}. \quad (14.42)$$

The first term in Equation (14.41) describes the common action on an oscillator of the additive and multiplicative forces while the second one is related to an oscillator response to the external periodic force. As one can see from (14.41), the first moment $\langle x \rangle$ shows non-monotonic dependence on the parameters of noise and periodic force. Moreover, for some values of these parameters an oscillator becomes unstable.

14.3 White noise

Equation (1.15) with $\eta = 0$ can be rewritten in the following form

$$\frac{d^2 x}{dt^2} + \gamma \frac{dx}{dt} + \omega^2 x = -\xi \frac{d^2 x}{dt^2} \quad (14.43)$$

Based on linear response theory, the output $x(t)$ of the system to the input $-\xi d^2 x/dt^2$ is

$$x(t) = \frac{1}{\omega_1} \int_0^t \exp \left[-\frac{\gamma}{2} \left(t - u \right) \right] \sin \left[\omega_1 \left(t - u \right) \right] \left(-\xi \frac{d^2 x(u)}{dt^2} \right) du \quad (14.44)$$

where $\omega_1 = \sqrt{\omega^2 - \gamma^2/4}$. Finding $d^2 x/dt^2$ from Equation (14.44), inserting it into Equation (14.43) and using the well-known formula for splitting the correlations,

$$\left\langle \xi(t) \xi(t_1) \frac{d^2 x}{dt}(t_1) \right\rangle = \left\langle \xi(t) \xi(t_1) \right\rangle \left\langle \frac{d^2 x}{dt}(t_1) \right\rangle \quad (14.45)$$

one obtains for white noise,

$$(1 - \gamma D) \frac{d^2}{dt^2} \langle x \rangle + \gamma \frac{d}{dt} \langle x \rangle + \omega^2 \langle x \rangle = 0 \quad (14.46)$$

which means the renormalization of oscillator's mass.

14.4   Symmetric dichotomous noise

Averaging Equation (12.46) over an ensemble of random functions $\xi(t)$ and using Equation (14.18) with $n = 1$ and $g = x$ leads to

$$\left(1 - \sigma^2\right) \frac{d^2 \langle x \rangle}{dt^2} + \left(\gamma \frac{d}{dt} + \omega^2\right) \langle x \rangle - \left(\gamma \frac{d}{dt} + \gamma\lambda + \omega^2\right) \langle \xi x \rangle = -R \quad (14.47)$$

where we assume white-noise correlations of noises $\xi(t)$ and $\eta(t)$, that is, $\langle \xi(t) \eta(t_1) \rangle = R\delta(t - t_1)$.

The new function $\langle \xi x \rangle$ enters Equation (14.47). One can obtain a second equation for the two functions $\langle x \rangle$ and $\langle \xi x \rangle$ by multiplying Equation (12.46) by $\xi(t)$ and averaging, using again Equation (14.18) with $g = x$ and with $g = dx/dt$,

$$\left[\left(1 - \sigma^2\right) \left(\frac{d}{dt} + \lambda\right)^2 + \gamma \frac{d}{dt} + \gamma\lambda + \omega^2\right] \langle \xi x \rangle - \sigma^2 \left(\gamma \frac{d}{dt} + \omega^2\right) \langle x \rangle = 0.$$

$$(14.48)$$

The use of dichotomous noise offers a major advantage over other types of colored noise by terminating an infinite set of higher-order correlations, using the fact that $\left\langle \xi(t)^2 \right\rangle = \sigma^2$. Eliminating $\langle \xi x \rangle$ from Equations (14.47) and (14.48), one obtains the following cumbersome equation for the first moment $\langle x(t) \rangle$

$$\left(1 - \sigma^2\right) \frac{d^4 \langle x \rangle}{dt^4} + \left[2\gamma + 2\lambda \left(1 - \sigma^2\right)\right] \frac{d^3 \langle x \rangle}{dt^3} +$$

$$\left[2\omega^2 + \gamma^2 + 3\gamma\lambda + \lambda^2 \left(1 - \sigma^2\right)\right] \frac{d^2 \langle x \rangle}{dt^2} + (\lambda + \gamma) \left(\gamma\lambda + 2\omega^2\right) \frac{d \langle x \rangle}{dt} +$$

$$\omega^2 \left[\omega^2 + \lambda(\lambda + \gamma)\right] \langle x \rangle = -R \left[\lambda^2 \left(1 - \sigma^2\right) + \gamma\lambda + \omega^2\right]. \quad (14.49)$$

Equations (14.47) and (14.48) can also be solved by the Laplace transform,

$$\langle x(p) \rangle = \int_0^\infty \langle x(t) \rangle \exp(-pt) \, dt, \quad (14.50)$$

with initial conditions $x(t = 0) = x_0$, $dx/dt \,(t = 0) = (dx/dt)_0$, $\langle \xi x \rangle$ $(t = 0) = \langle \xi y \rangle \,(t = 0) = 0$. Then, for $R = 0$, one obtains,

$$\langle x(p) \rangle = \frac{\left[M - \sigma^2 p + \lambda^2\right] \left[px_0 + (dx/dt)_0\right] + 2\gamma M x_0}{M \left(p^2 + 2p\gamma + \omega^2\right) - \sigma^2 p^2 \left(p + \lambda\right)^2} \quad (14.51)$$

where $M = (p + \lambda)^2 + 2\gamma(p + \lambda) + \omega^2$.

Another way to analyze the differential Equation (1.25) (with $\eta = 0$) is to transform it into an integro-differential Equation [165]. For this purpose, we rewrite this equation in the following form

$$L\{x\} = -\xi \frac{d^2 x}{dt^2}; \qquad L\{x\} \equiv \frac{d^2}{dt^2} + 2\gamma \frac{d}{dt} + \omega^2. \qquad (14.52)$$

Applying the operator L^{-1} to the first of Eqs. (14.52), one obtains

$$x(t) = -L^{-1}\left\{\xi \frac{d^2 x}{dt^2}\right\}. \qquad (14.53)$$

Using the fact that $L\{L^{-1}\{f\}\} = f$, one can easily check that the integral operator L^{-1} inverse to the differential operator L, defined in (14.52), has the following form

$$L^{-1}\{\psi(t)\} = \frac{1}{\omega_1}\int_0^t dt_1 \exp\left[-\gamma(t-t_1)\right]\sin\left[\omega_1(t-t_1)\right]\psi(t_1); \quad (14.54)$$

$$\omega_1 = \sqrt{\omega^2 - \gamma^2/4} \qquad (14.55)$$

Using (14.53) and (14.54) yields

$$x(t) = -\frac{1}{\omega_1}\int_0^t dt_1 \exp\left[-\gamma(t-t_1)\right]\sin\left[\omega_1(t-t_1)\right]\xi(t_1)\frac{d^2 x}{dt^2}(t_1) \quad (14.56)$$

and

$$\frac{d^2 x}{dt^2} = \xi(t)\frac{d^2 x}{dt^2} - \frac{1}{\omega_1}\int_0^t dt_1\{(\omega^2 - 2\gamma^2)\sin\left[\omega_1(t-t_1)\right] +$$

$$2\omega_1\gamma\cos\left[\omega_1(t-t_1)\right]\}\exp\left[-\gamma(t-t_1)\right]\xi(t_1)\frac{d^2 x}{dt^2}(t_1). \qquad (14.57)$$

Multiplying Equation (14.57) by $\xi(t)$, substituting the resulting expression into Equation (14.52), and averaging over noise, one obtains

$$\frac{d^2}{dt^2}\langle x\rangle + 2\gamma\frac{d}{dt}\langle x\rangle + \omega^2\langle x\rangle = \langle\xi^2(t)(d^2 x/dt^2)\rangle -$$

$$\frac{1}{\omega_1}\int_0^t dt_1\{(\omega^2 - 2\gamma^2)\sin\left[\omega_1(t-t_1)\right] +$$

$$2\omega_1\gamma\cos\left[\omega_1(t-t_1)\right]\}\exp\left[-\gamma(t-t_1)\right]\langle\xi(t)\xi(t_1)\rangle\frac{d^2}{dt^2}\langle x\rangle(t_1). \quad (14.58)$$

For the analysis of Equation (14.58), we use Equation (14.45), which gives

$$\frac{d^2}{dt^2}\langle x\rangle + 2\gamma\frac{d}{dt}\langle x\rangle + \omega^2\langle x\rangle = \sigma^2\frac{d^2}{dt^2}\langle x\rangle -$$

$$\frac{\sigma^2}{\omega_1}\int_0^t dt_1\{(\omega^2 - 2\gamma^2)\sin\left[\omega_1(t-t_1)\right] +$$

$$2\omega_1\gamma\cos\left[\omega_1(t-t_1)\right]\}\exp\left[-(\lambda+\gamma)(t-t_1)\right](d^2\langle x\rangle/dt^2)(t_1). \quad (14.59)$$

The integrand in Equation (14.59) has the form $f(t_1) g(t - t_1)$. Using the convolution theorem for the Laplace transform [173],

$$L\left\{\int_0^t f(t_1) g(t - t_1) dt_1\right\} = L\{f\} L\{g\}, \qquad (14.60)$$

one gets Equation (14.49), obtained earlier by the other method.

14.5 Asymmetric dichotomous noise

Starting from Equation (1.15), using Equation (14.18) with $n = 2$, and averaging over noise yields

$$\left[(1 + \sigma^2) \frac{d^2}{dt^2} + \gamma \frac{d}{dt} + \omega^2\right] \langle x \rangle + \Delta \left(\frac{d}{dt} + \lambda\right)^2 \langle \xi x \rangle = 0. \qquad (14.61)$$

A second equation for the two functions $\langle x \rangle$ and $\langle \xi x \rangle$ can be obtained from Equation (1.24) by using Equation (14.18) (with $n = 2$ and $n = 1$),

$$\left[T \frac{d^2}{dt^2} \langle x \rangle + (1 + \sigma^2) \left(\gamma \frac{d}{dt} + \omega^2\right)\right] \langle x \rangle -$$

$$\left[\Delta^3 \left(\frac{d}{dt} + \lambda\right)^2 + \Delta\gamma \left(\frac{d}{dt} + \lambda\right) + \Delta\omega^2\right] \langle \xi x \rangle = 0. \qquad (14.62)$$

Excluding the correlator $\langle \xi x \rangle$ from Equations (14.61) and (14.62), one obtains a following fourth-order differential equation for $\langle x \rangle$,

$$\left[(1 + \sigma^2)^2 + \Delta^2\right] \frac{d^4}{dt^4} \langle x \rangle + [\Delta^2 (\gamma + 2\lambda) + 2\lambda (1 + \sigma^2) +$$

$$2\gamma (1 + \sigma^2)] \frac{d^3}{dt^3} \langle x \rangle + [\Delta^2 (\lambda + \gamma) + (1 + \sigma^2) (2\omega^2 + \lambda + 3\gamma) +$$

$$\omega^2 \Delta^2 + \gamma^2] \frac{d^2}{dt^2} \langle x \rangle + [\lambda (1 + \sigma^2 + \Delta^2) + \gamma] (2\omega^2 + \lambda\gamma) \frac{d}{dt} \langle x \rangle +$$

$$\omega^2 \left[\omega^2 + \lambda\gamma + \lambda^2 (1 + \sigma^2 + \Delta^2)\right] \langle x \rangle = 0 \qquad (14.63)$$

14.6 Second moment

For calculation of the second moment $\langle x^2 \rangle$, let us multiply the first of Equations (14.16) by $2x$ and the second by $2y$ yields

$$\frac{d}{dt} x^2 = 2xy; \quad \frac{d}{dt} y^2 + \xi \frac{dy^2}{dt} + 2\gamma y^2 + 2\omega^2 xy = 2y\eta(t) + 2Ay \sin(\Omega t) \quad (14.64)$$

Using the well-known formula for splitting the correlations, which is exact for the Ornstein-Uhlenbeck noise (1.20),

$$\langle \xi(t) \eta(t) y \rangle = \langle \xi(t) \eta(t) \rangle \langle y \rangle. \tag{14.65}$$

and averaging Equation (14.64) by using (14.18) yields

$$\frac{d}{dt} \langle x^2 \rangle = 2 \langle xy \rangle; \quad \left(\frac{d}{dt} + 2\gamma \right) \langle y^2 \rangle +$$

$$\left(\frac{d}{dt} + \frac{1}{\tau} \right) \langle \xi y^2 \rangle + 2\omega^2 \langle xy \rangle = 2\varepsilon + 2A \langle y \rangle \sin(\Omega t). \tag{14.66}$$

Analogously, multiplying Equations (14.16) by y and x, respectively, and summing leads to

$$\frac{d}{dt} xy = y^2 - \xi x \left(-\xi \frac{dy}{dt} - \gamma y - \omega^2 x + \eta(t) \right) - \gamma xy - \omega^2 x^2 +$$

$$\eta x + 2Ax \sin(\Omega t) = y^2 + \sigma^2 \left[\frac{d}{dt} (xy) - y^2 \right] + \gamma \xi xy + \omega^2 \xi x^2 -$$

$$\xi \eta x - \gamma xy - \omega^2 x^2 + x\eta + 2Ax \sin(\Omega t) \tag{14.67}$$

which yields after averaging and using $\langle x\xi\eta \rangle = 0$ yields for $\sigma^2 < 1$,

$$\frac{d}{dt} \langle xy \rangle = \langle y^2 \rangle + \gamma \langle \xi xy \rangle + \omega^2 \langle \xi x^2 \rangle - \gamma \langle xy \rangle - \omega^2 \langle x^2 \rangle + 2A \langle x \rangle \sin(\Omega t). \tag{14.68}$$

Equations (14.66) and (14.68) contain new correlators $\langle \xi x^2 \rangle$, $\langle \xi y^2 \rangle$ and $\langle \xi xy \rangle$. One can calculate these correlators by multiplying Equations (14.16) and (14.67) by $2\xi x, 2\xi y$ and ξ, respectively, and averaging,

$$\left(\frac{d}{dt} + \frac{1}{\tau} \right) \langle \xi x^2 \rangle = 2 \langle \xi xy \rangle \tag{14.69}$$

$$\left(\frac{d}{dt} + \frac{1}{\tau} + 2\gamma \right) \langle \xi y^2 \rangle + 2\omega^2 \langle \xi xy \rangle = 2A\kappa \langle y \rangle \sin(\Omega t) \tag{14.70}$$

$$\left(\frac{d}{dt} + \frac{1}{\tau} + \gamma \right) \langle \xi xy \rangle = \langle \xi y^2 \rangle - \omega^2 \langle \xi x^2 \rangle + \omega^2 \sigma^2 \langle x^2 \rangle + \lambda \langle x \rangle + 2A \langle \xi x \rangle \sin(\Omega t). \tag{14.71}$$

By this means we obtain six equations, (14.66) and (14.68)-(14.71), for the six variables $\langle x^2 \rangle$, $\langle y^2 \rangle$, $\langle xy \rangle$, $\langle \xi x^2 \rangle$, $\langle \xi y^2 \rangle$, and $\langle \xi xy \rangle$. We will not write here the cumbersome dynamic equations for the second moments, which can be easily obtained from this system of differential equations, but shall restrict our attention to the case of small fluctuations of mass ($\sigma^2 < 1$) in the absence of an external periodic force, $A = 0$ and in the limiting case of the long time limit, $t \to \infty$,

$$\left\langle x^2 \right\rangle = \frac{\alpha}{2\gamma\omega^2} \frac{1}{1 - \sigma^2\beta/\gamma} \tag{14.72}$$

where

$$\beta = \frac{\omega^2 + \gamma \left(1 + 2\gamma\tau\right) \left(\gamma + 2\omega^2\tau\right)}{2\omega^2\tau + \left(2\omega^2\tau + \gamma + 1/\tau\right)\left(1 + 2\gamma\tau\right)}. \tag{14.73}$$

For white noise, Equation (14.72) reduces to

$$\left\langle x^2 \right\rangle = \frac{\alpha}{2\gamma\omega^2}. \tag{14.74}$$

This result coincides with the well-known result for "free" Brownian motion with $\omega^2 = 0$. For free Brownian particle, $\omega^2 \to 0$ and one obtains $\left\langle x^2 \right\rangle \to \infty$, as it should be for Brownian motion. The independence of the stationary results on the mass fluctuation is due to the fact that the multiplicative random force appears in Equation (1.18) in front of the higher derivative. It is remarkable that these results are significantly different from the stationary second moments for the white-noise random frequency $\left\langle x^2 \right\rangle_\omega$ and random damping $\left\langle x^2 \right\rangle_\gamma$,

$$\left\langle x^2 \right\rangle_\omega = \frac{\alpha}{2\omega^2\left(\gamma - D\omega^2\right)} : \quad \left\langle x^2 \right\rangle_\gamma = \frac{\alpha}{2\gamma\omega^2(1 - 2\gamma D)} \tag{14.75}$$

showing the "energetic" instability [14]. It turns out that for symmetric dichotomous noise, the stationary second moment $\left\langle x^2 \right\rangle$ for the mass fluctuations, in contrast to its white noise form (14.74), may lead to instability, $\left\langle x^2 \right\rangle < 0$ for $\sigma^2\beta > \gamma$.

The correlation function can be found along the same lines as was done for the second moment by multiplying Equations (14.16) by $x(t_1)$ and averaging the resulting equations, which gives

$$\frac{d}{dt}\left\langle x(t_1) x(t)\right\rangle = \left\langle x(t_1) y(t)\right\rangle \tag{14.76}$$

$$\frac{d}{dt}\left\langle x(t_1) y(t)\right\rangle = -\left\langle \xi(t) \frac{d}{dt}\left\langle x(t_1) y(t)\right\rangle\right\rangle - \gamma\left\langle x(t_1) y(t)\right\rangle - \omega^2\left\langle x(t_1) x(t)\right\rangle.$$

The new correlator $\left\langle \xi(t) \left\langle x(t_1) y(t)\right\rangle\right\rangle$ can be found by using Equation (14.45) leading to

$$\left(\frac{d}{dt} + \lambda\right)\left\langle \xi(t) x(t_1) x(t)\right\rangle = \left\langle \xi(t) x(t_1) y(t)\right\rangle;$$

$$\left(\frac{d}{dt} + \lambda\right)^2 \left\langle \xi(t) x(t_1) x(t)\right\rangle + \sigma^2 \frac{d}{dt}\left\langle x(t_1) y(t)\right\rangle + \gamma\left\langle \xi(t) x(t_1) y(t)\right\rangle +$$

$$\omega^2 \left\langle \xi(t) x(t_1) x(t)\right\rangle = 0. \tag{14.77}$$

From Equations (14.76)-(14.77), one can find the fourth-order differential equation for the correlation function $\langle x(t_1) x(t) \rangle$, which, due to the linearity of this equation, coincides with Equation (14.49) for the first moment.

For dichotomous noise, the correlation function shows a non-monotonic dependence on both the noise strength σ^2 and the inverse correlation time λ^{-1}.

14.7 Stochastic resonance in the oscillator with random mass

In the broad sense, stochastic resonance means the non-monotonic dependence of the output signal or some function of it, as a function of some characteristic of noise or of the periodic signal. Stochastic resonance is a phenomenon found in a dynamic nonlinear system driven by a combination of a random and a periodic force [68], [186]. However, it was shown [73], [14] that stochastic resonance also occurs in a linear system subject to multiplicative color noise. The system described by Equation (1.19), does fall into this category. Rewriting Equation (1.23) as two first-order differential equations

$$\frac{dx}{dt} = y; \qquad [1 + \sigma^2 + \Delta \xi(t)] \frac{dy}{dt} + \gamma y + \omega^2 x = a \sin(\Omega t) \qquad (14.78)$$

and averaging over the random noise, one obtains

$$\frac{d\langle x \rangle}{dt} = \langle y \rangle; \qquad \frac{d\langle y \rangle}{dt} + \left[(1 + \sigma^2) \frac{d}{dt} + \gamma \right] \langle y \rangle +$$

$$\Delta \left(\frac{d}{dt} + \lambda \right) \langle \xi y \rangle + \omega^2 \langle x \rangle = a \sin(\Omega t). \qquad (14.79)$$

To split the correlator $\langle \xi dy/dt \rangle$ we use the well-known Shapiro-Loginov procedure [29] which yields for exponentially correlated noise

$$\left\langle \xi \frac{dg}{dt} \right\rangle = \left(\frac{d}{dt} + \lambda \right) \langle \xi g \rangle. \qquad (14.80)$$

Multiplying Equations (14.78) by $\xi(t)$ and averaging yields

$$\left(\frac{d}{dt} + \lambda \right) \langle \xi x \rangle = \langle \xi y \rangle;$$

$$\left[(1 + \sigma^2 + \Delta^2) \left(\frac{d}{dt} + \lambda \right) + \gamma \right] \langle \xi y \rangle + \Delta \sigma^2 \frac{d}{dt} \langle y \rangle + \omega^2 \langle \xi x \rangle = 0.$$

$$(14.81)$$

Averaging Equations (14.79)-(14.81) and using Equations (14.80) yields the following four equations for the four functions, $\langle x \rangle$, $\langle y \rangle$, $\langle \xi x \rangle$ and $\langle \xi y \rangle$,

$$\frac{d}{dt}\langle x \rangle = \langle y \rangle; \qquad \left[(1+\sigma)\frac{d}{dt}+\gamma\right]\langle y \rangle + \Delta\left(\frac{d}{dt}+\lambda\right)\langle \xi y \rangle + \omega^2\langle x \rangle = 0;$$

$$\left(\frac{d}{dt}+\lambda\right)\langle \xi x \rangle = \langle \xi y \rangle;$$

$$\left[(1+\sigma+\Delta^2)\left(\frac{d}{dt}+\lambda\right)+\gamma\right]\langle \xi y \rangle + \Delta\sigma\frac{d}{dt}\langle y \rangle + \omega^2\langle \xi x \rangle = \langle \xi\eta \rangle.$$

$$\tag{14.82}$$

From Equations (14.82), one gets a fourth-order differential equation for $\langle x \rangle$,

$$\left(f_1 f_2 - \Delta^2\sigma^2\right)\frac{d^4}{dt^4}\langle x \rangle + \left(\gamma f_2 + f_1 f_3 - 2\Delta^2\lambda\sigma^2\right)\frac{d^3}{dt^3}\langle x \rangle +$$

$$\left(\omega^2 f_2 + f_1 f_4 + \gamma f_3 - \sigma^2\Delta^2\lambda^2\right)\frac{d^2}{dt^2}\langle x \rangle + \left(\gamma f_4 + \omega^2 f_3\right)\frac{d}{dt}\langle x \rangle +$$

$$\omega^2 f_4\langle x \rangle = \left(\omega^2 f_2 + \omega^2 - f_2\Omega^2\right)a\sin(\Omega t) + (2\lambda f_2 + \gamma)a\Omega\cos(\Omega t) \tag{14.83}$$

where

$$f_1 = 1+\sigma^2; \quad f_2 = 1+\sigma^2+\Delta^2; \quad f_3 = \gamma+2\lambda f_2; \quad f_4 = \omega^2 + \lambda(\gamma+\lambda f_2).$$

$$\tag{14.84}$$

In a similar way, one can obtain the equation for the second moment $\langle x^2 \rangle$, associated with Equation (1.23), which is transformed into six equations for six variables, $\langle x^2 \rangle$, $\langle y^2 \rangle$, $\langle xy \rangle$, $\langle \xi x^2 \rangle$, $\langle \xi y^2 \rangle$ and $\langle \xi xy \rangle$, but we shall not write down these cumbersome equations.

Analogous to the cases of random frequency and random damping [105], we seek the solution of Equation (14.83) in the form

$$\langle x \rangle = A\sin(\Omega t + \phi). \tag{14.85}$$

One easily finds

$$A = \left(\frac{f_5^2 + f_6^2}{f_7^2 + f_8^2}\right)^{1/2}; \qquad \phi = \tan^{-1}\left(\frac{f_5 f_7 + f_6 f_8}{f_5 f_8 - f_6 f_7}\right) \tag{14.86}$$

with

$$f_5 = (f_4 - f_2\Omega^2)a; \qquad f_6 = \Omega f_3 a; \tag{14.87}$$

$$f_7 = \Omega^3\left(\gamma f_2 + f_1 f_3 - 2\Delta^2\lambda\sigma^2\right) - \Omega\left(\gamma f_4 + \omega^2 f_3\right) \tag{14.88}$$

$$f_8 = \omega^2 f_4 - \Omega^2\left(\omega^2 f_2 + f_1 f_4 + \gamma f_3 - \Delta^2\lambda^2\sigma^2\right) + \Omega^4\left(f_1 f_2 - \Delta^2\sigma^2\right)$$

where the functions $f_1 ... f_4$ were defined in Equation (12.21).

One can compare Equations (14.85)-(14.88) with the equations for the first moment $\langle x \rangle$, obtained [105] for the cases of random frequency and random damping, respectively, subject to a symmetric dichotomous noise, and extended afterwards [187], [188] to the case of an asymmetric noise. All these equation are of a fourth order with the same dependence on the frequency Ω of the external field but a slightly different dependence of the parameters of the noise.

The amplitude A of the output signal depends on the characteristics σ, Δ, λ of the asymmetric dichotomous noise and the frequency Ω of the input signal. The signal-to-noise ratio is of frequent use in the analysis of stochastic resonance, which involves the use of the second moments. For simplicity,

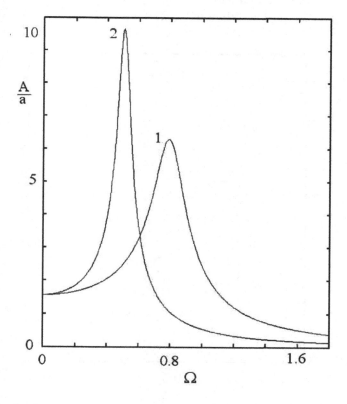

Fig. 14.1 Stochastic resonance. Comparison of Output-Input ratio as a function of the external frequency Ω, showing the increase of the output signal in the presense of noise (graph 2 with $\sigma^2 = 1, 5$, $\Delta = 0.2$ and $\lambda = 0.5$) compared to the output signal without noise (graph 1 with $\sigma^2 = \Delta = \lambda = 0$). Parameters: $\omega = 0, 8$ and $\gamma = 0.5$.

we call stochastic resonance the non-monotonic behavior of the ratio A/a of the amplitude of the output signal A to the amplitude a of the input signal. (Output-Input ratio, OIR). Figure 14.1, which shows the dependence of the OIR on the external frequency confirms the existence of the phenomenon of stochastic resonance. Moreover, the presence of noise, which usually plays the destructive role, results here in an increase of the output signal, thereby improving the efficiency of a system in the amplification of a weak signal. In the absence of noise, the usual dynamic resonance occurs, when the frequency of an external force approaches the eigenfrequency of an oscillator.

Figures 14.2 and 14.3 show the Ω - dependence of the OIR for parameters $\gamma = \lambda = \sigma^2 = \Delta = 1$ and different eigenfrequencies $\omega < 1$ (Figure 14.2) and $\omega > 1$ (Figure 14.3). The values of the maxima increase with a decrease of

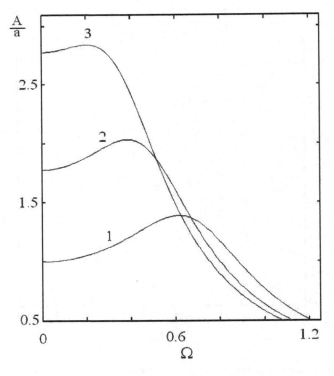

Fig. 14.2   Output-Input ratio as the function of the external frequency Ω for different $\omega < 1$ and $\gamma = \lambda = \sigma^2 = \Delta = 1$. Curves 1, 2 and 3 correspond to $\omega = 1.0$, $\omega = 0.75$ and $\omega = 0.6$, respectively.

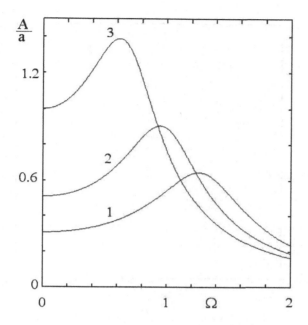

Fig. 14.3 Output-Input ratio as the function of the external frequency Ω for different $\omega > 1$ and $\gamma = \lambda = \sigma^2 = \Delta = 1$. Curves 1,2 and 3 correspond to $\omega = 1.8$, $\omega = 1.4$ and $\omega = 1.0$, respectively.

ω on both plots, although the positions of maxima are shifted to the right with the decrease of ω for $\omega > 1$ and to the left for $\omega < 1$. The resonant dependence of OIR as the function of the inverse correlation time λ, the strength σ^2 and the asymmetry of noise Δ is shown in Figures 14.4-14.6.

14.8 Stability conditions

Here we consider the more complicated problem of the stability of the solutions. For a deterministic equation, the stability of the fixed points is defined by the sign of α, found from the solution of the form $\exp(\alpha t)$ of a linearized equation near the fixed points. The situation is quite different for a stochastic equation. The first moment $\langle x(t) \rangle$ and higher moments become unstable for some values of the parameters. However, the usual linear stability analysis, which leads to instability thresholds, turns out to be different for different moments making them unsuitable for a stability

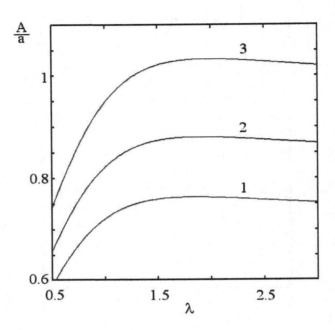

Fig. 14.4   Output-Input ratio as the function of the inverse correlation time λ for $\gamma = 0.05$, $\Omega = 1$, $\sigma^2 = 2.8$ and $\Delta = 2.5$. Curves 1, 2 and 3 correspond to $\omega = 1.85$, $\omega = 1.8$ and $\omega = 1.75$, respectively.

analysis. A rigorous mathematical analysis of random dynamic systems shows [179] that, similar to the order–deterministic chaos transition in nonlinear deterministic equations, the stability of a stochastic differential equation is defined by the sign of Lyapunov exponents λ. This means that for stability analysis, one has to go from the Langevin-type equations to the associated Fokker-Planck equations which describe the properties of statistical ensembles and to calculate the Lyapunov index λ, defined by [179]

$$\lambda = \frac{1}{2} \left\langle \frac{\partial \ln\left(x^2\right)}{\partial t} \right\rangle = \left\langle \frac{\partial x/\partial t}{x} \right\rangle. \tag{14.89}$$

One can see from Equation (14.89) that it is convenient to replace the variable x in the Langevin equations with the variable $z = (dx/dt)/x$,

$$\frac{dz}{d\tau} = \frac{d^2x/d\tau^2}{x} - \frac{(dx/d\tau)^2}{x^2} \equiv \frac{d^2x/d\tau^2}{x} - z^2. \tag{14.90}$$

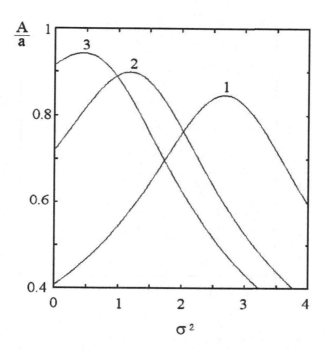

Fig. 14.5   Output-Input ratio as the function of the noise strength σ^2 for $\gamma = \lambda = \sigma^2 = \Delta = 1$. Curves 1, 2 and 3 correspond to $\omega = 1.85$, $\omega = 1.8$ and $\omega = 1.75$, respectively.

The Lyapunov index λ now takes the following form [189]

$$\lambda = \int_{-\infty}^{\infty} z P_{st}(z) dz \qquad (14.91)$$

where $P_{st}(z)$ is the stationary solution of the Fokker-Planck equations corresponded to the Langevin equations expressing in the variable z.

Replacing the variable x in Equation (14.90) by the variable z leads to

$$\frac{dz}{d\tau} = A(z) + \xi_1 B(z) \qquad (14.92)$$

where

$$A(z) = -z^2 - B(z); \qquad B(z) = \frac{1}{R}\left(1 + \sigma^2\right)\left(\gamma z + \omega^2\right); \qquad (14.93)$$

$$\xi_1(t) = \frac{\Delta}{1 + \sigma^2}\xi(t).$$

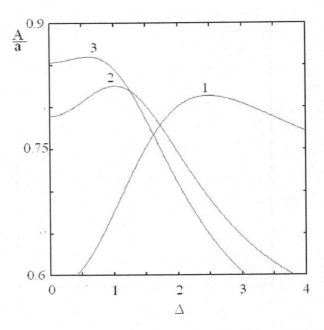

Fig. 14.6 Output-Input rate as the function of the noise asymmetry Δ for $\gamma = \lambda = \Delta = 1$ and $\sigma^2 = 2.5$. curves 1, 2 and 3 correspond to $\omega = 1.45$, $\omega = 1.65$, and $\omega = 1.7$, respectively.

14.9 White noise

Although for the foregoing reasons white noise cannot be used for an oscillator with random mass, this approximation might be useful for similar problems. For white noise,

$$\langle \xi_1 (t_1) \xi_1 (t_2) \rangle = \frac{D\Delta^2}{(1 + \sigma^2)^2} \delta (t_1 - t_2). \tag{14.94}$$

The Fokker-Planck equation associated with Equation (14.94) has the following form (Stratonovich interpretation) [34]

$$\frac{\partial P (z, \tau)}{\partial \tau} = -\frac{\partial}{\partial z} [A (z) P] + \frac{D\Delta^2}{2 (1 + \sigma^2)^2} \frac{\partial}{\partial z} B (z) \left[\frac{\partial}{\partial z} B (z) P \right]. \tag{14.95}$$

For the stationary case, this reduces to

$$- [A (z) P_{st}] + \frac{D\Delta^2}{2 (1 + \sigma^2)^2} B (z) \frac{\partial}{\partial z} [B (z) P_{st}] = J \tag{14.96}$$

where J is the constant probability current.

The solution of the homogeneous Equation (14.96) (with $J = 0$) is

$$P_{st}(z) = \frac{C}{B(z)} \exp\left[\frac{2\left(1+\sigma^2\right)^2}{D\Delta^2} \int_c^z dy \frac{A(y)}{B^2(y)}\right]. \tag{14.97}$$

The solution of the inhomogeneous Equation (14.96) can be obtained by the method of variation of constants,

$$P_{sr}(z) = \frac{4J\left(1+\sigma^2\right)^2}{D\Delta^2 B(z)} \exp\left[\frac{2\left(1+\sigma^2\right)^2}{D\Delta^2} \int_c^z \frac{A(x)}{B^2(x)} dx\right] \times$$

$$\int_c^z \frac{dy}{B(y)} \exp\left[-\frac{2\left(1+\sigma^2\right)^2}{D\Delta^2} \int_c^y dx \frac{A(x)}{B^2(x)}\right] +$$

$$\frac{C}{B(z)} \exp\left(\frac{2\left(1+\sigma^2\right)^2}{D\Delta^2} \int_c^z dy \frac{A(y)}{B^2(y)}\right). \tag{14.98}$$

The constant C and the reference point $z = c$ are not important for our analysis, and we may set $C = 0$ for $c = -\infty$.

Inserting (14.93) into (14.98) transforms Equation (14.98) into the following form,

$$P_{st} = P_{sr}(w) = \frac{J\omega_1}{N\Delta\gamma_1}(w)^{-c-1} \exp\left[-g\left(w - \frac{1}{w}\right)\right] \int_{-\infty}^w dx\, x^{c-1} \exp\left[g\left(x - \frac{1}{x}\right)\right] \tag{14.99}$$

where

$$c = \frac{4\gamma\left(1+\sigma^2\right)^2}{D\Delta^2}. \tag{14.100}$$

There is no need to perform an analysis of Equation (14.99), because the analogous calculation has been performed for the case of random damping [189] yielding the following result after substitution in Equation (14.91),

$$\lambda = \frac{4\int_0^\infty du\, K_1\left(8b\sinh u\right)\sinh\left[(1-c)\,u\right]}{\pi^2\left[J_{\gamma/D}^2\left(4/D\right) + Y_{\gamma/D}^2\left(4/D\right)\right]} \tag{14.101}$$

where K_1 is a modified Bessel function of the second kind, and J and Y are Bessel functions of the first and second kind, respectively. The Bessel functions are always positive, and the sign of the Lyapunov exponent λ is the same as the sign of the hyperbolic function $\sinh\left[(1-c)\,u\right]$, which is the sign of $1 - c$.

Therefore, an oscillator with fluctuating mass becomes unstable when $c < 1$, i.e., the instability of the fixed point $x = 0$ occurs for $D > 4\left(1+\sigma^2\right)^2\gamma/\Delta^2$ and does not depend on the oscillator frequency ω.

It is instructive to compare these results with those obtained for additive noise and two other multiplicative sources of noise acting on the frequency and on the damping coefficient. The first moment is not changed for the case of additive noise and changed only slightly for the case of random frequency, replacing the original frequency ω by a renormalized frequency $\sqrt{\omega^2 - \gamma^2/4}$. However, for random damping coefficient γ excited by white noise of strength D, the randomness results in the replacement of γ by $\gamma(1 - \gamma D)$. In the case of fluctuating mass, for the case considered of fast oscillations $(\omega > \gamma)$, randomness leads to the renormalization of the frequency ω and (small) renormalization of the damping coefficient γ by $\omega\sqrt{1 - \gamma D}$ and $\gamma(1 - \gamma D)$, respectively. The statistical analysis of Lyapunov exponents for fast oscillations and white noise shows that the instability occurs for large strength of noise.

Another peculiar feature of the Lyapunov index (14.91), which is defined by the stationary distribution function (14.99), is the change of its sign as a function of the oscillator and noise parameters. The integrals in (14.91) and (14.99) contain the multi-valued function x^c. Therefore, the transformed integration contour in the complex x-plane starts at $x = \infty$, makes the branch cut at $x = 0$, encircling the origin and returning to its starting point. The integrals of these type appear in the theory of Bessel functions, namely [190]

$$J_\nu(w) = -\frac{\exp(i\nu\pi)}{2\pi i}\left(\frac{w}{2}\right)^\nu \int dx\, x^{\nu-1}\exp\left[-x + \frac{w^2}{x}\right]. \qquad (14.102)$$

Finally, one obtains from a comparison of (14.102) with (14.99),

$$\lambda \approx \frac{J_c(g)}{J_{c-1}(g)}. \qquad (14.103)$$

If the parameter c, defined in (14.100), is an integer, then the well-known oscillating behavior of the Bessel functions of integer order [190] leads one to conclude that with the change of parameter g, the Lyapunov index changes its sign showing that with changing of the noise strength, there are many noise-induced reentrant transitions from the ordered to disordered states, and *vice versa*.

14.10   Dichotomous noise

According to [95], the stationary solution of the Fokker-Planck equation, corresponding to the Langevin Equation (14.92) having the correlation function (12.52), has the following form

$$P_{st}\left(z\right)=N\frac{B}{\sigma^2 B^2 - A^2}\exp\left[-\frac{1}{2\tau}\int^z dx\left[\frac{1}{A\left(x\right)-\sigma B\left(x\right)}+\frac{1}{A\left(x\right)+\sigma B\left(x\right)}\right]\right].$$

(14.104)

Equation (14.104) has been analyzed for different forms of functions $A\left(x\right)$ and $B\left(x\right)$: $A = -x$, $B = 1$ [191]; $A = x$, $B = -x$ [192] ; $A = x - x^m$, $B = x$, [193]; $A = x - x^3$, $B = 1$ [194]; $A = x^3$. $B = x$ [195], [196]; $A = x - x^2$, $B = x$ [95].

The zeroes of functions $F_{\pm}\left(x\right) = \pm\sigma B\left(x\right) - A\left(x\right)$ determine the boundary of $P_{st}\left(z\right)$, which diverges or vanishes at the boundaries and determine the boundary of support of $P_{st}\left(z\right)$. The latter means that a system will approach the state z located in intervals (z_2, z_1) or (z_4, z_3), depending on its initial position. Another important characteristic of $P_{st}\left(z\right)$ is the location of its extrema, which define the macroscopic steady states. The steady states x_m of (14.104) obey the following Equation [95]

$$Ax_m - \sigma^2\tau B\left(x_m\right)\frac{d}{dx}A\left(x_m\right) + 2\tau A\left(x_m\right)\frac{d}{dx}A\left(x_m\right) - \tau\frac{A\left(x_m\right)^2 B\left(x_m\right)}{dB\left(x_m\right)/dx}.$$

(14.105)

The first term in (14.105) defines the deterministic steady states whereas the next two terms relate to the white-noise limit ($\sigma \to \infty$, $\tau \to 0$ with $\sigma^2\tau = const$). Finally, the last two terms define the corrections coming from the final correlation time τ.

For

$$A = \alpha x^2 + \beta x + \kappa; \qquad B = \beta x + \kappa \qquad (14.106)$$

with

$$\alpha = -1;\ \beta = -\frac{\gamma}{R}\left(1 + \sigma^2\right);\ \kappa = -\frac{\omega^2}{R}\left(1 + \sigma^2\right) \qquad (14.107)$$

one obtains, according to (14.93),

$$\sigma^2 B^2 - A^2 = \left[\alpha x^2 + (1 + \sigma)(\beta x + \kappa)\right]\left[-\alpha x^2 - (1 - \sigma)(\beta x + \kappa)\right]$$

(14.108)

and

$$\frac{1}{2\tau}\int^z\frac{dx}{\alpha x^2 + (1+\sigma)(\beta x + \kappa)} = -\frac{1}{2\tau\alpha(x_1 - x_2)}\left[\int^z\frac{dx}{(x-x_1)} - \int^z\frac{dx}{(x-x_2)}\right] =$$

$$-\frac{1}{2\tau\alpha(x_1 - x_2)}\ln\frac{z - x_1}{(z - x_2)} \qquad (14.109)$$

$$\frac{1}{2\tau}\int\limits^{z}\frac{dx}{-\alpha x^2-(1-\sigma)\,(\beta x+\kappa)}=\frac{1}{2\tau\alpha\,(x_3-x_4)}\left[\int\limits^{z}\frac{dx}{(x-x_3)}-\int\limits^{z}\frac{dx}{(x-x_4)}\right]=$$

$$\frac{1}{2\tau\alpha\,(x_3-x_4)}\ln\frac{z-x_3}{(z-x_4)}\qquad(14.110)$$

where

$$x_{1,2}=-\frac{(1+\sigma)\,\beta}{2\alpha}\pm\sqrt{\left(\frac{(1+\sigma)\,\beta}{2\alpha}\right)^2-\frac{(1+\sigma)\,\kappa}{\alpha}}$$

$$=-\gamma Q_+\pm\sqrt{\gamma^2 Q_+^2-\omega^2 Q_+}\qquad(14.111)$$

$$x_{3,4}=-\frac{(1-\sigma)\,\beta}{2\alpha}\pm\sqrt{\left(\frac{(1-\sigma)\,\beta}{2\alpha}\right)^2+\frac{(1-\sigma)\,\kappa}{\alpha}}$$

$$=-\gamma Q_-\pm\sqrt{\gamma^2 Q_-^2-\omega^2 Q_-}\qquad(14.112)$$

with

$$Q_\pm=\frac{\left(1+\sigma^2\right)(1\pm\sigma)}{R}.\qquad(14.113)$$

Inserting (14.109)-(14.113) into (14.104) gives

$$P_{st}\,(z)=N\,(z-x_1)^{-1-[2\tau\alpha(x_1-x_2)]^{-1}}\,(z-x_2)^{-1+[2\tau\alpha(x_1-x_2)]^{-1}}\times$$

$$(z-x_3)^{-1+[2\tau(x_3-x_4)]^{-1}}\,(z-x_4)^{-1-[2\tau(x_3-x_4)]^{-1}}.\qquad(14.114)$$

According to (14.91), this defines the boundary of stability of the fixed point $x=0$ for different values of parameters $\gamma Q_\pm$ and $\omega^2 Q_\pm$, which depend on characteristics ω^2, γ of an oscillator and σ, Δ and τ of the noise.

14.11   Resonance phenomena

The simplest example of mechanical resonance is a harmonic oscillator subject to a periodic force, where the steady-state amplitude of the oscillator approaches infinity when the external force frequency approaches the eigenfrequency of the oscillator. This phenomenon was probably already known to the ancient Egyptians who invented the water clock, but the classical demonstration of dynamic resonance are quite recent architectural flaws uncovered in the US. The first was the Takoma bridge which was destroyed by the wind force at the resonance frequency, and the second was the Paramount Communication Building in New York where the winds twisted the top floors and pried windows loose from their casements.

The well-known phenomena of deterministic chaos, stochastic and vibrational resonances occur for an oscillator with random mass if one adds one or two periodic forces to the oscillator equation. Stochastic resonance manifests itself in the fact that the noise, which always plays a distractive role, appears as a constructive force, increasing the output signal as a function of noise intensity. Like stochastic resonance, vibrational resonance manifests itself in the enhancement of a weak periodic signal through a high-frequency periodic field, instead of through noise as in the case of stochastic resonance.

One of greatest achievements of twentieth-century physics was establishing a deep relationship between deterministic and random phenomena. The widely studied phenomena of "deterministic chaos" and "stochastic resonance" might sound contradictory, consisting of half-deterministic and half-random terms. In addition to stochastic resonance, another exciting phenomenon is deterministic chaos which appears in equations without any random force. Deterministic chaos means an exponential divergence in time of the solutions for even the smallest change in the initial conditions. Therefore, there exists a close connection between determinism and randomness, even they are apparently different forms of behavior [63].

Here we consider a new manifestation of the resonance of an oscillator. The dynamic equation of motion of a bistable underdamped one-dimensional oscillator driven by a multiplicative random force $\alpha\xi(t)$, an additive random force $\beta\eta(t)$, and two periodic forces, $A\sin(\omega t)$ and $C\sin(\Omega t)$, has the following form

$$\frac{d^2x}{dt^2} + \gamma\frac{dx}{dt} - \omega_0^2 x + \alpha\xi(t)\,x + bx^3 = \beta\eta(t) + A\sin(\omega t) + C\sin(\Omega t). \quad (14.115)$$

The dynamic resonance mentioned above corresponds to $\gamma = b = \alpha = \beta = C = 0$ and $\omega \to \omega_0$. Let us consider some other limiting cases of Equation (14.115).

1) Brownian motion ($\omega_0 = b = A = C = 0$) has been studied most widely with many applications. The equilibrium distribution comes from the balance of two contrary processes: the random force which tends to increase the velocity of the Brownian particle and the damped force which tries to stop the particle [14].

2) The double-well oscillator with additive noise ($\alpha = A = C = 0$) and small damping, $\gamma << \omega$, shows two or three peaks in the power spectrum (Fourier component of the correlation function) descriptive of fluctuation

transitions between the two stable points of the potential, small intra-well vibrations and the over-the-barrier vibrations [147].

3) Stochastic resonance (SR) in overdamped $(d^2x/dt^2 = \alpha = C = 0)$ and underdamped $(\alpha = C = 0)$ oscillators is a very interesting and counterintuitive phenomenon, where the noise increases a weak input signal. SR occurs in the case that a deterministic time-scale of the external periodic field is synchronized with a stochastic time-scale, determined by the Kramers transition rate over the barrier.

4) Stochastic resonance in a linear overdamped oscillator $(d^2x/dt^2 = \beta = b = C = 0)$, as distinct from the nonlinear case, allows an exact solution [73]-[74]. However, this effect occurs only when the multiplicative noise $\xi(t)$ is colored and not white.

5) Vibrational resonance $(\alpha = \beta = 0)$, which occurs in a deterministic system, manifests itself in the enhancement of a weak periodic signal through a high-frequency periodic field, instead of through noise, as in the case of stochastic resonance.

6) "Erratic" behavior shows up as a "random-like" phenomenon in a simple system $(d^2x/dt^2 = \alpha = \beta = 0)$ with two incommensurate external frequencies, ω and Ω.

14.11.1 *Vibrational resonance*

Analysis of the stochastic resonance in an oscillator with fluctuating mass has been performed in Chapter 5. Like stochastic resonance, vibrational resonance manifests itself in the enhancement of a weak periodic signal through a high-frequency periodic field, instead of through noise as in the case of stochastic resonance. The deterministic equation of motion then has the following form,

$$\frac{d^2x}{dt^2} + \gamma\frac{dx}{dt} - \omega_0^2 x + \beta x^3 = A\sin(\omega t) + C\sin(\Omega t). \qquad (14.116)$$

Equation (14.116) describes an oscillator moving in a symmetric double-well potential $V(x) = -\omega_0^2 x^2/2 + \beta x^4/4$ with a maximum at $x^* = 0$ and two minima $x_\pm$ with the depth d of the wells,

$$x_\pm^* = \pm\sqrt{\frac{\omega_0^2}{\beta}} \qquad d = \frac{\omega_0^4}{4\beta}. \qquad (14.117)$$

The amplitude of the output signal as a function of the amplitude C of the high-frequency field has a bell shape, showing the phenomenon of vibrational resonance. For ω close to the frequency ω_0 of the free oscillations, there are two resonance peaks, whereas for smaller ω, there is only one resonance peak. These different results correspond to two different oscillatory processes, jumps between the two wells and oscillations inside one well.

Assuming that $\Omega >> \omega$, resonance-like behavior ("vibrational resonance" [156]) manifests itself in the response of the system at the low-frequency ω, which depends on the amplitude C and the frequency Ω of the high-frequency signal. The latter plays a role similar to that of noise in SR. If the amplitude C is larger than the barrier height d, the field during each half period π/Ω transfers the system from one potential well to the other. Moreover, the two frequencies ω and Ω are similar to the frequencies of the periodic signal and the Kramers rate of jumps between the two minima of the underdamped oscillator. Therefore, by choosing an appropriate relation between the input signal $A \sin(\omega t)$ and the amplitude C of the large signal (or the strength of the noise) one can obtain a non-monotonic dependence of the output signal on the amplitude C (vibration resonance) or on the noise strength (stochastic resonance). To put this another way [197], both noise in SR and the high-frequency signal in vibrational resonance change the parameters of the system response to a low-frequency signal.

Let us now pass to an approximate analytical solution of Equation (14.116). In accordance with the two times scales in this equation, we seek a solution of Equation (14.116) in the form

$$x(t) = y(t) - \frac{C \sin(\Omega t)}{\Omega^2} \qquad (14.118)$$

where the first term varies significantly only over times t, while the second term varies much more rapidly. On substituting Equation (14.118) into (14.116), one can average over a single cycle of $\sin(\Omega t)$. Then, odd powers of $\sin(\Omega t)$ vanish upon averaging, while the $\sin^2(\Omega t)$ term gives $1/2$. In this way, one obtains the following equation for $y(t)$,

$$\frac{d^2 y}{dt^2} + \gamma \frac{dy}{dt} - \left(\omega_0^2 - \frac{3bC^2}{2\Omega^4}\right) y + by^3 = A \sin \omega t \qquad (14.119)$$

with

$$y_0^* = 0; \qquad y_\pm^* = \pm\sqrt{\frac{\omega_0^2 - 3bC^2/2\Omega^4}{b}} \; ; \qquad d = \frac{\left[\omega_0^2 - 3bC^2/2\Omega^4\right]^4}{4b} .$$
$$(14.120)$$

One can say that Equation (14.119) is the "coarse-grained" version (with respect to time) of Equation (14.116). For $3\beta C^2/2\Omega^4 > \omega_0^2$, the phenomenon of dynamic stabilization [198] occurs, namely, the high-frequency external field transforms the previously unstable position $\Psi = 0$ into a stable position.

Seeking the solution of Equation (14.119) of the form

$$y(t) \approx y^* + \Theta \sin(\omega t - \theta) \tag{14.121}$$

and linearizing Equation (14.119) in Θ gives

$$\Theta = \frac{A}{\sqrt{\left(\omega_1^2 - \omega^2\right)^2 + \gamma^2 \omega^2}} \tag{14.122}$$

where

$$\omega_1^2 = \frac{3bC^2}{2\Omega^4} - \omega_0^2 + 3b\left(y^*\right)^2. \tag{14.123}$$

A resonance in the linearized Equation (14.119) occurs when $\omega_1 = \omega$, which, after substituting in Equation (14.119), leads to the following relations between the amplitudes and frequencies of the two driving fields which produce the resonant behavior,

$$\omega^2 = \frac{3bC^2}{2\Omega^4} - \omega_0^2 + \frac{3bA^2}{4\gamma^2 \omega^2}. \tag{14.124}$$

In addition to the resonance phenomenon, one can study [158] the influence of the positions and depths of the potential on the vibrational resonance. Assuming that $\omega_0^2 = b$, which means, according to Equation (14.120), that positions of minima remain fixed, let us raise the question for which value of a control parameter C the ratio of the output signal Θ to the input signal A is maximal. According to Equation (14.122), this occurs when $S = \left(\omega_1^2 - \omega^2\right)^2 + \gamma^2 \omega^2$ is minimal, which is determined by the condition $dS/dC = 0$, which, using (14.123) with $\omega_0^2 = b$, results in

$$2\left(\omega_1^2 - \omega^2\right)\frac{d\omega_1^2}{dC} = \frac{3bC}{\Omega^4}\left[\frac{3bC^2}{2\Omega^4} - b - \omega^2 + 3b\left(y^*\right)^2\right] = 0 \tag{14.125}$$

or, for $y_0^* = 0$,

$$C^2 = \frac{2\Omega^4}{3b}\left(b + \omega^2\right) \tag{14.126}$$

and for $y_\pm^* = \pm\sqrt{\frac{\omega_0^2}{b} - \frac{3C^2}{2\Omega^4}}$,

$$C^2 = \frac{\Omega^4}{3b}\left(2b - \omega^2\right). \tag{14.127}$$

Equation (14.127) has real solutions for C only if $2b > \omega^2$.

Thus far, we considered equal values of two control parameters, $\omega_0^2 = b$ changing the depths of potential and keeping the positions of minima $x_\pm^*$ unaltered. Analogously, one can assume that $\omega_0^4 = b$ changing thereby the distance between minima and not the potential depth. Then, one obtains, for $y_0^* = 0$,

$$C^2 = \frac{2\Omega^2}{3}\left(2 + \frac{\omega^2}{b}\right) \tag{14.128}$$

and for $y_\pm^* = \pm\sqrt{\frac{\omega_0^2}{b} - \frac{3C^2}{2\Omega^4}}$,

$$C^2 = \frac{2\Omega^2}{3}\left(2 - \frac{\omega^2}{b}\right) \tag{14.129}$$

with the proviso that $2b > \omega^2$.

All the above results have been obtained for an underdamped oscillator. It turns out [157], [199] that a similar effect also takes place for an over-damped oscillator ($d^2x/dt^2 = 0$ in Equation (14.116)). The influence of the additional additive noise on the vibrational resonance, and the advantages of the vibrational resonance compared to the stochastic resonance in the detection of weak signals have been studied [200].

For an oscillator with random mass one has to perform the preceding analysis of Equation (14.116), based on dividing its solution in the two time scales (Equation (14.118)) followed by the linearization of Equation (14.119) for the slowly changing solution. The subsequent analysis of an oscillator equation with one periodic force is quite analogous to analysis of Equation (1.23), which describes the stochastic resonance phenomenon.

Equation (14.116) describes an oscillator moving in symmetric double-well potential. The vibrational resonance in the quintic oscillator with the potential of the form

$$V(x) = \frac{1}{2}\omega_0^2 x^2 + \frac{1}{4}bx^4 + \frac{1}{6}cx^6 \tag{14.130}$$

was studied in [201]. Finally, the vibrational resonance and an appearance of chaos in the Van der Pol oscillator were investigated in [202]. Because of the many applications in physics, chemistry, biology and engineering, vibrational resonance still attracts great interest, and new applications will surely be found in the future.

14.11.2 Deterministic chaos ("Erratic" behavior)

One of the great achievements of twentieth-century physics was the prediction of deterministic chaos which appears in the equations without any random force [203]. Deterministic chaos means an exponential increase in time of the solutions for even the smallest change in the initial conditions. Therefore, to obtain a "deterministic" solution, one would have to specify the initial conditions to an infinite number of digits. Otherwise, the solutions of deterministic equations show chaotic behavior. Deterministic chaos occurs if the differential equations are nonlinear and contain at least three variables. This points to the important difference between underdamped and overdamped equations of an oscillator, since deterministic chaos may occur only in the underdamped oscillator. Here, we present an example of "erratic" behavior, which, like deterministic chaos, is drawn midway between deterministic and stochastic behavior.

Consider the simple example of an overdamped oscillator subject to two periodic fields,

$$\frac{dx}{dt} + \omega^2 x = C_1 \cos(\omega_1 t) + C_2 \cos(\omega_2 t). \tag{14.131}$$

We show that the solutions of this equation are "erratic", being intermediate between deterministic and chaotic solutions.

The stationary solutions of Equation (14.131) have the following form

$$x(t) = \frac{C_1}{\omega_1} \sin(\omega_1 t) + \frac{C_2}{\omega_2} \sin(\omega_2 t). \tag{14.132}$$

Replace the continuous time in Equation (14.131) by discrete times $2\pi n/\omega_2$ [204]. The solution of this equation then becomes

$$x\left(n\frac{2\pi}{\omega_2}\right) = x(0) + \frac{C_1}{\omega_1}\sin\left(2\pi n\frac{\omega_1}{\omega_2}\right). \tag{14.133}$$

If ω_1/ω_2 is an irrational number, the sin factor in (14.133) will never vanish and the motion becomes "erratic". The properties of "erratic" motion can be understood from the analysis of the correlation function associated with the n-th and $(n+m)$-th points,

$$C(2\pi m\omega_1/\omega_2) = \lim_{N\to\infty} \frac{1}{N} \sum_{n=0}^{N} x(2\pi n\omega_1/\omega_2) \, x[2\pi(n+m)\omega_1/\omega_2]$$

$$= x^2(0) + x(0)(C_1/\omega_1) \lim_{N\to\infty} \frac{1}{N} \sum_{n=0}^{N} \{\sin(2\pi n\omega_1/\omega_2) +$$

$$\sin[2\pi(n+m)\omega_1/\omega_2] + (C_1/\omega_1)^2 \lim_{N\to\infty} \frac{1}{N} \sum_{n=0}^{N} \times$$

$$\sin(2\pi n\omega_1/\omega_2) \sin[2\pi(n+m)\omega_1/\omega_2]. \tag{14.134}$$

Using the well-known relations between the trigonometric functions, one obtains

$$C\left(m\frac{2\pi\omega_1}{\omega_2}\right) = x^2(0) + \frac{1}{2}\left(\frac{C_1}{\omega_1}\right)^2 \cos\left(m\frac{2\pi\omega_1}{\omega_2}\right). \qquad (14.135)$$

The Fourier spectrum of the correlation function (14.135) depends on the ratio ω_1/ω_2. If this ratio is a rational number, this spectrum will contain a finite number of peaks. However, for irrational ω_1/ω_2, the spectrum becomes broadband, what is typical of deterministic chaos. However, this "erratic" behavior arises from a simple "integrable" Equation (14.131), which distinguishes it from deterministic chaos.

Chapter 15

In the Future...

One hundred years have passed since the explanation of the phenomenon of Brownian motion by Einstein, Smoluchowski and Langevin. Like other great inventions such as general relativity, a need was generated for a special mathematical description. It has taken the new concept of stochastic differential equations to express the molecular-kinetic aspect of this phenomenon. The fluctuation force exerted on the Brownian particle by the molecules of a surrounding medium was depicted by an additive noise in the differential equation of motion of a Brownian particle. During the last hundred years, a large body of work has been devoted to the modelling of different phenomena in physics, chemistry, biology and sociology through the use of additive and multiplicative noise in differential equations. This noise has an internal or an external origin, respectively. A dynamic oscillator is the simplest toy model for different phenomena in Nature, and taking into account the surrounding noise makes these models more adequate. Many new phenomena have been found during the past hundred years, and their number is growing like a snowball. An example is Brownian motion with adhesion, where the surrounding molecules not only collide with the Brownian particle inducing a zigzag motion, but also adhere to it for a random period of time, thereby increasing the mass of the Brownian particle. Due to many applications in physics, chemistry, biology and engineering, the model of an oscillator with random mass will find many applications in the future.The most impressive discovery is the deep relationship between determinism and stochasticity - their general meaning, like that of relativity and quantum mechanics, is certainly going beyond the scope of physics. The widely studied phenomena of "deterministic chaos" [205] and "stochastic resonance" (SR) [132] might sound internally contradictory, consisting of half-deterministic and half-random terms. In fact, deterministic chaos denotes a random type of behavior in deterministic systems, while SR shows deterministic-like behavior in random systems. These peculiar features show that determinism and randomness are complementary, rather than contradictory, phenomena [206]. The peculiarity of SR lies in the fact that noise, which usually appears as a destructive factor, may play a constructive role. The reader will certainly enjoy (together with the author) the title of a recent article [207] "Noise is good for the brain". Many surprises are still in store in stochastic phenomena in Nature...

Bibliography

[1] M. Moshinski and Yu. F. Smirnov, *The Harmonic Oscillator in Modern Physics* [Harvard, The Netherlands, 1996].

[2] A. Einstein in R. H. Furth ed. *Investigations on the Theory of the Brownian Motion* [Dover, New York, 1954].

[3] M. Smoluchowski, Ann. Phys., **21**, 756 (1906).

[4] P. Langevin, Compt. Rend. **146**, 530, (1908).

[5] L. D. Landau and E. M. Lifshitz, *Statistical Physics,* [Pergamon 1980].

[6] A. H. Nayfeh, *Perturbation Methods* [Wiley, New York 1974].

[7] S. C. Venkataramani, T. M. Antonsen, Jr., E. Ott, amd J. C. Sommerer, Physica D **96**, 66 (1996).

[8] R. Graham, M. Hoihenerbach, and A. Schenze, Phys. Rev. Lett., **48**, 1396 (1982).

[9] W. Kohler and G. C. Papanicolaou, in *Springer Lectures in Physics,* **70** [Springer, New York 1977].

[10] V. L. Klyatskin and V. I. Tatarskii, Soviet Physics - Usp. **16**, 494 (1974).

[11] M. Turelli, *Theoretical Population Biology* [Academic, New York, 1977].

[12] H. Takayasu, A-H. Sato, and M. Takayasu, Phys. Rev. Lett., **79**, 966 (1997).

[13] V. I. Klyatskin, *Stochastic Equations and Waves in Randomly Inhomogeneous Media,* [Nauka,Moscow, 1980] (In Russian); V. I. Klyatskin, Sov. Phys - Acoustic **26**, 113 (1980).

[14] M. Gitterman, Physica A **352**, 305 (2005).

[15] B. West and V. Seshadri, J. Geophys. Res. **86**, 4293 (1981).

[16] M. Gitterman, Phys. Rev. E **70**, 036116 (2004).

[17] A. Onuki, J. Phys. Condens. Matter **9**, 6119 (1997).

[18] J. M. Chomaz and A. Couairon, Phys. Fluids, **11**, 2977 (1999).

[19] T. E. Faber, *Fluid Dynamics for Physicists* [Cambridge University Press, Cambridge, 1995].

[20] F. Heslot and A. Libchaber, Phys. Scr. **T9**, 126 (1985).

[21] A. Saul and K. Showalter, *Ocsillations and Travel Waves in Chemical Systemns,* ed. R.J. Field and M. Burger [Wiley, New York, 1985].

[22] M. Gitterman, B. Ya. Shapiro, and I. Shapiro, Phys. Rev. E **65**, 174510 (2002).

[23] M. Gitterman, Phys. Rev. E **69**, 041101 (2004).

[24] M. Gitterman, J. Phys. C, 248, 012049 (2010); M. Gitterman, J. Stat. Phys. **146**, 239 (2010); M. Gitterman and V. I. Kljatskin, Phys. Rev. E, **81**, 051139 (2010); M. Gitterman and I. Shapiro, J. Stat. Phys. **144**, 139 (2011); M. Gitterman, J. Mod.Phys. **2**, 1136 (2010); M. Gitterman, World J. Mech, **2**, 113 (2012).

[25] J. Luczka, P. Hanggi and A. Gadomski, Phys. Rev. E **51**, 5762 (1995).

[26] M. Sewbawe Abdalla, Phys. Rev. A **34**, 4598 (1986).

[27] J. Portman, M. Khasin, S. W. Shaw, and M. I. Dykman, Bulletin of the APS, March Meeting, 2010.

[28] K. Furutsu, J. Res. NBS, D **67**, 303 (1963); E.A. Novikov, Sov. Phys.-JETP, **20**, 1290 (1965).

[29] V. E. Shapiro and V. M. Loginov, Physica A **91**, 563 (1978).

[30] R. Mankin, K. Laas, and T. Laas, Phys. Rev. E **78**, 031120 (2008); K. Laas, R. Mankin, and A. Rekker, Phys. Rev. E **79**, 051128 (2009).

[31] Zhang Lu, Zhong Su-Chuan, Peng Hao, and Luo Mao-Kang, Chin. Phys. Lett., 28, 090505 (2011).

[32] P. Hanggi, Z. Phys. B **36**, 271 (1980).

[33] E. Nelson, *Dynamical Theories of Brownian Motion* [Princeton, 1967].

[34] N. G. van Kampen, *Stochastic Processes in Physics and Chemistry* [North-Holland, Amsterdam 1992].

[35] Yu. L. Klimontovich, Physica A **163**, 515 (1992)

[36] J. J. Brey, J. M. Casado, and M. Morillo, Z. Phys. B. **66**, 263 (1987).

[37] Romanczuk, P.; Bèar, M.; Ebeling, W.; Lindner, B.; Schimansky-Geier, L. Europ. Phys. J., 202,. 1 (2012).

[38] U. Erdmann, W. Ebeling, F. Schweitzer, and L. Schimansky-Geier, Eur. Phys. J. B, **15**, 105 (2000).

[39] P. Jung and P. Hanggi, Phys. Rev. A **35**, 4464 (1987).

[40] P. Hanggi and P. Jung, Adv. Chem. Phys. **89**, 239 (1995).

[41] S. Z. Ke, D. J. Wu, and L. Cao, Eur. Phys. J. B **12**, 119 (1999).

[42] J-P. Bouchaud and A. Georges, Phys. Rep., **195**, 127 (1990).

[43] R. Meltzer and J. Klafter, Phys. Rep., **333**, 1 (2000).

[44] S. I. Denisov and W. Horsthemke, Phys. Rev. E **62**, 7729 (2000).

[45] F. Lillo and R. N. Mantegna, Phys. Rev. E **61**, R4675 (2000).

[46] R. Rozenfeld, J. Luczka, P. Talkner, Phys. Lett.A, **249**, 409 (1998).

[47] M. Gitterman and V. Steinberg, Phys. Fluids, **23**, 2154 (1980).

[48] R. Kubo, Rep. Prog. Phys. **29**, 255 (1966).

[49] H. Xia, O. Ishibara, and A. Hirose, Phys. Fluids, B **5**, 2892 (1993).

[50] M. W. Reeks, Phys. Fluids, **31**, 1314 (1988).

[51] M. Gitterman, Phys. Rev. E **52**, 303 (1995); Physica A **221**, 330 (1995).

[52] E. Fermi, Phys. Rev. **75**, 1169 (1949).

[53] C. Van den Broeck and R. Kawai, Phys. Rev. E **57**, 3866 (1998).

[54] V. E. Shapiro, Phys. Rev. E **48**, 109 (1993).

[55] L. Landau and E. Lifshitz, *Mechanics* [Pergamon, 1976].

[56] A. Fulinski and P.F. Gora, Phys. Rev. E **48**, 3510 (1993).

[57] F. Liu, B. Hu, and W. Wang, Phys. Rev. E **63**, 031907 (2001).

[58] C. Zhou, J. Kurtis, I. Z. Kiss, and J. L. Hudson, Phys. Rev. Lett. **89**, 014101 (2002).

[59] B. Allen and J. D. Romano, Phys. Rev. D **59**, 102001 (1999).

[60] Y. El-Mohri, L. E. Antonuk, Q. Zhao, M. Maolinbay, X. Rong, K.-W. Jee, S. Nassif, and C. Cionka, Med. Phys.**27**, 1855 (2000).

[61] K. S. Fa, Chem. Phys. **287**, 1 (2003)

[62] M. R. Young and S. Singh, Phys. Rev. A **38**, 238 (1988).

[63] M. Gitterman, J. Phys. A **32**, L293 (1999).

[64] V. Berdichevsky and M. Gitterman, Phys. Rev. E **60**, 1494 (1999).

[65] P. Li, L. R. Nie, C. Z. Shu, S. Hu, Q. Shao, J. Stat. Phys. **146**, 1184 (2012).

[66] Li Jinghui, Commun. Theor. Phys. **50**, 1159, (2008).

[67] Li Dong-Sheng and Li Jing-Hui, Commun. Theor. Phys. **53**, 298 (2006).

[68] R. Benzi, S. Sutera, and A. Vulpiani, J. Phys. A **14**, L453 (1981); C. Nicolis, Tellus **34**, 1 (1982).

[69] S. Sinha, Physica A **270**, 204 (1999).

[70] H. Gang, T. Ditzinger, C.Z. Ning, and H. Haken, Phys. Rev. Lett. **71**, 807 (1993); J.L. Cabrera, J. Corronogoitia, and F.J. de la Rubia, Phys. Rev. Lett. **82**, 2816 (1999).

[71] J.J. Collins, C.C. Chow, A.C. Capela, and T.T.Imhoff, Phys. Rev. E **54**, 5575 (1996).

[72] C.R. Doerind and J.C. Gadoua, Phys. Rev. Lett. **69**, 2318 (1992).

[73] A. Fulinski, Phys. Rev. E **52**, 4523 (1995).

[74] V. Berdichevsky and M. Gitterman, Europhys. Lett.**36**, 161 (1996).

[75] A.V. Bazikin and K. Seki, Europhys. Lett. **40**, 117 (1997).

[76] V. Bezak, Czech. J. Phys. **48**, 529 (1998).

[77] A.V. Bazikin and K. Seki, and F. Shibata, Phys. Rev. E **57**, 6555 (1998)

[78] A.A. Zaikin, J. Kurths, and L. Schimansky-Geier, Phys. Rev. Lett. **85**, 227 (2000); A.A. Zaikin, K. Murali, and J. Kurths, Phys. Rev. E **63**, 020103(R) (2001); Jing-Hui Li, Europhys. Lett. **82**, 50006 (2008).

[79] Jing-Hui Li, Phys. Rev. E **76**, 021113 (2007).

[80] J.M.G. Vilar and J.M. Rubi, Phys. Rev. Lett. **78**, 2886 (1997); Jing-Hui Li and Yin-Xia Han, Commun. Theor. Phys. **48**, 605 (2007).

[81] A.S. Pikovksy and J. Kurths, Phys. Rev. Lett. **78**, 775 (1997); C. Palenzuela, R. Toral, C.R. Mirasso, O. Calvo, and J.D. Gunton, Europhys. Lett. **56**, 347 (2001).

[82] I. Goychuk and P. Hanggi, Phys. Rev. E **59**, 5137 (1999).

[83] L. Gammaitoni, M. Locher, A. Bulsara, P. Hänggi, J. Neff, K. Wiesenfeld, W. Ditto, and M.E. Inchiosa, Phys. Rev. Lett. **82**, 4574 (1999); M. Locher, M.E. Inchiosa, J. Neff, A. Bulasra, K. Wiesenfeld, L. Gammaitoni, P. Hanggi, W. Ditto, Phys. Rev. E **62**, 317 (2000); J.F. Lindner, J. Mason, J. Neff, B.J. Breen, W.L. Ditto, and A.R. Bulsara, Phys. Rev. E **63**, 041107 (2001).

[84] G.S. Jeon and M.Y. Choi, Phys. Rev. B **66**, 064514 (2002); J.L. Cabrera, J. Corronogoitia, and F.J. de la Rubia, Phys. Rev. Lett. **82**, 2816 (1999).

[85] J.J. Collins, C.C. Chow, A.C. Capela, and T.T. Imhoff, Phys. Rev. E **54**,

5575 (1996); K. Park, Ying-Cheng Lai, Zong-Hua Liu, and A. Nachman, Phys. Lett. A **326**, 391(2004).

[86] A. Fuliinski and P. F. Gora, J. Stat. Phys. **101**, 483 (2000).

[87] M. I. Dykman, D. G. Luchinsky, P. V. E. McClintock, N. D. Stein, and N. G. Stocks, Phys. Rev. A **46**, R1713 (1992).

[88] L. Cao and D. J. Wu, Europhys. Lett. **61**, 593 (2003).

[89] B.-Q. Ai, X.-J. Wang, G.-T. Liu, and L.-G. Liu, Phys. Rev. E **67**, 022903 (2003).

[90] W. Horsthemke and M. Malek-Mansour, Z. Phys. B **24**, 307 (1976).

[91] L. Arnold and W. Horsthemke, and R. Lefewer, Z. Phys. B **29**, 867 (1978).

[92] W. Horsthemke, and R. Lefewer, *Noise Induced Phase Transitions* [Springer, Berlin 1984].

[93] A. V. Soldatov, Mod. Phys. Lett. **7**, 1253 (1993).

[94] C. Van den Broeck and P. Hanggi, Phys. Rev. A **30**, 2730 (1983).

[95] K. Kitahara,W. Horsthemke, R. Lefever, and Y. Inara, Prog. Theor. Phys. **64**, 1233 (1980).

[96] J. M. Sancho, M. San Miguel, L. Pesquera, and M. A. Rodriquez, Physica A **142**, 532 (1987).

[97] F. J. de la Rubia, Phys. Lett. A **110**, 17 (1985).

[98] I. Dayan, M. Gitterman, and G. H. Weiss, Phys. Rev. A **46**, 757 (1992).

[99] R. N. Mantegna and B. Spagnolo, Phys. Rev. Lett. **76**, 563 (1996).

[100] N. V. Agudov, A. A. Dubkov, and B. Spagnolo, Physica A **325**, 144 (2003).

[101] J. Mills, Phys. Lett. A **133**, 295 (2000).

[102] M. Gitterman, J. Phys A **34**, L355 (2001).

[103] V. Berdichevsky and M. Gitterman, J. Phys. A **31**, 9773 (1998).

[104] V. Berdichevsky and M. Gitterman, Phys. Rev. E **59**, R9 (1999).

[105] M. Gitterman, Phys. Rev. E **65**, 031103 (2002).

[106] R. Graham and A. Schenze, Phys. Rev. A **25**, 1731 (1982).

[107] R. Graham, Phys. Rev. A **25**, 3234 (1982).

[108] R. Graham, Phys. Rev. A **10**, 1762 (1974).

[109] H. Brand and A. Schenze, Phys. Rev. A **20**, 1628 (1979).

[110] S Kabashima and T. Kawakubo, in *Systems Far From Equilibrium*, ed. L. Garrido, Lecture Notes in Physics **132** [Spriger, Berlin 1980].

[111] D.-J. Wu, L. Cao, and S.-Z. Ke, Phys. Rev. E **50**, 2496 (1994).

[112] J. M. Sancho, M. San Miguel, S. L. Katz, and J. D. Gunton, Phys. Rev. A **26**, 1589 (1982).

[113] Ya Lia and J.-R. Li, Phys. Rev. E **53**, 5786 (1996).

[114] D. Mei, C. Xie, and L. Zhang, Phys. Rev. E **68**, 051102 (2003).

[115] E. Cortes and K. Lindenberg, Physica A **123**, 99 (1984).

[116] A. Teubel, U. Behn, and A. Kuhnel, Z Phys. B **71**, 392 (1988).

[117] I. S. Gradstein and I. M. Ryzhik, *Tables of Integrals,Series and Products*, [Academic, Boston 1994].

[118] E. T. Whittaker and G. N. Watson, *A Course of Modern Analysis* [Cambridge University Press, 1027].

[119] P. Jung, Z. Phys. B **76**, 521 (1989).

[120] H. Risken, *The Fokker-Planck Equation* [Springer, Berlin 1996].

[121] F. Bass and M. Gitterman, Radiophys. Radioastron **61**, 71 (2001).

[122] J. Masoliver, B. J. West, and K. Lindenberg, Phys. Rev. A **35**, 3086 (1987).

[123] Ya Lia and J.-R. Li, Phys. Rev. E **53**, 5764 (1996).

[124] A.J. R. Madureira, P. Hanggi, V. Buonomano, and W. A. Rodrigez Jr., Phys. Rev. E **51**, 3849 (1995).

[125] Y. Jin and W. Xu, Chaos, Soliton, Fractals **23**, 275 (2005).

[126] L. Gammaitoni, P. Hanggi, P. Jung, and F. Marchesoni, Rev. Mod. Phys. **70**, 223 (1998).

[127] T. Iwai, Physica A **300**, 350 (2001).

[128] D. Dan and A. M. Jayannavar, Physica A **345**, 404 (2005).

[129] P. Jung, Phys. Rep. **234**, 175 (1993)

[130] F. Moss, *Contemporary problems in statistical physics*, ed. G. H. Weiss [SIAM Philadelphia 1994]. p.205.

[131] B. McNamara and K. Wiesenfeld, Phys. Rev. A **39**, 4854 (1988).

[132] L. Gammaitoni, P. Hanggi, P. Jung, and F. Marchesoni, Rev. Mod. Phys. **70**, 233 (1998).

[133] Y. Jia, S.-N. Yu, and J-R. Li, Phys. Rev. E **62**, 1869 (2000).

[134] Y. Jia, X.-p. Zheng, X.-m. Hu, and J-r. Li, Phys. Rev. E **63**, 031107 (2001).

[135] X. Luo and S. Zhu, Phys. Rev. E **67**, 021104 (2003).

[136] R. Rozenfeld, A. Neiman, and L. Schimansky-Geier, Phys. Rev. E **62**, 051107 (2001).

[137] J.-H. Li, Phys. Rev. E **66**, 031104 (2002).

[138] E. Guardia and M. San Miguel, Phys. Lett. A **109**, 9 (1985).

[139] P. Yung and F. Marcheconi, Chaos, **21**, 047516 (2011).

[140] P. Pechukas and P. Hanggi, Phys. Rev. Lett. **73**, 2772 (1994).

[141] J. H. Li, D. Y. Xing, J. M. Dond, and B. Hu, Phys. Rev. E **60**, 1324 (1999).

[142] C. Van den Broeck, Phys. Rev. E **47**, 4579 (1993).

[143] J. C. Fletcher, S. Havlin, and G. H. Weiss, J. Stat. Phys. **51**, 215 (1988).

[144] M. Gitterman and G. H. Weiss, J. Stat. Phys. **74**, 941 (1988).

[145] J. J. Brey and J. Casado- Pascual, Physica A **212**, 123 (1994).

[146] J. M. Porra, Phys. Rev. E **55**, 6533 (1997).

[147] M. I. Dykman, D. G. Luchinsky, R. Mannela, P. V. E. Nc Clintock, N. D. Stein, and N. G. Stocks, Il Nuovo Cimento D **17**, 661 (1995).

[148] L. H'walisz, P. Jung, P. Hanggi, P. Talkner, and L. Shimansky-Geier, Z. Phys. B **77**, 471 (1989).

[149] K.-G. Wang and J. Masoliver, Physica A **231**, 615 (1996).

[150] M. C. Wang and G. E. Uhlenbeck, Rev. Mod. Phys. **17**, 323 (1945).

[151] J. J. Stoker, *Nonlinear vibrations* [Interscience, New York 1950].

[152] M. Gitterman, R. I. Shrager, and G. H. Weiss, Phys. Lett. A **142**, 84 (1989).

[153] A. R. Bulsara, K. Lindenberg, and K. E. Shuler, J. Stat. Phys. **27**, 787 (1982).

[154] A. B. Budgor, K. Lindenberg, and K. E. Shuler, J. Stat. Phys. **15**, 375, (1976).

[155] M. I. Dykman and M. A. Krivoglaz, Physica A **104**, 495 (1980); M. I. Dykman, S. M. Soskin, and M. A. Krivoglaz, Physica A **133**, 53 (1985);

M. I. Dykman, R. Mannela, PVE. McClintock, F. Moss, and S. M. Soskin, Phys. Rev. A **37**, 1303 (1988).

[156] P. S. Landa and P. V. E. McClintock, J. Phys. A **33**, L433 (2000).

[157] I. I. Blekhman and P. S. Landa, Int. J. Nonlin. Mech. **39**, 421 (2004).

[158] Rajasekar S., Jeyakumari S., Chinnathambi V., Sanuan M. A., J. Phys A **43**, 465101 (2010).

[159] V. N. Chizhevsky, E. Smeu, and G. Giacomelli, Phys. Rev. Lett. **91**, 220602 (2003).

[160] A. A. Zaikin, L. Lopez, J. P. Baltanas, J. Kurth, and M. A. F. Sanjuan, Phys. Rev. E **66**, 011106 (2002).

[161] F. Guo, Y. R. Zhou, S. Q. Jiang, and T. X. Gu, J Phys. A **39**, 13861 (2006).

[162] K. Lindenberg, V. Seshadri, and B. West, Physica A **105**, 445 (1988).

[163] S. K. Banik and D. S. Ray, J. Phys. A **31**, 3937 (1998).

[164] B. V. Bobryk and A. Chreszczyk, Physica A **315**, 225 (2002).

[165] R. C. Bourret, H. Frish, and A. Pouquet, Physica **65**, 303 (1973).

[166] M. A. Leibowitz, J. Math. Phys. **4**, 852 (1963).

[167] J. Barre and T. Dauxaois, Europhys. Lett. **55**, 164 (2001).

[168] I. M. Lifshitz, S. A. Gredeskul, and L. A. Pastur, *Introduction to the Theory of Disorder Systems*, [Wiley, New York 1988].

[169] M. F. Dimentberg, *Nonlinear Stochastic Problems of Mechanicasl Oscillations*, [Nauka, Moscow 1980], in Russians.

[170] Li-Juan Ning, Xu Wei, and Yao Ming-Li, Chin. Phys B, **17**, 486 (2008).

[171] Li-Juan Ning and Xu Wei, Chin J. Phys. **46**, 611 (2008).

[172] Li-Juan Ning, Xu Wei, and Yao Ming-Li, Chin. Phys. **16**, 2595 (2007).

[173] S. Roul, "Stable Polynomials." §10.13 in *Programming for Mathematicians*, [Springer-Verlag, Berlin 2000].

[174] V. Mendez, W. Hosthemke, P. Mestres, and D. Campos, Phys. Rev. E **84**, 041137 (2011).

[175] J. Luczka, J. Phys. A **21**, 3063 (1988); J. Luczka and J. Stadkowski, Czech. J. Phys. B **39**, 689. (1989).

[176] M. Gitterman, *The Noisy Oscillator: the First Hundred Years, from Einstein, Until Now* [World Scientific, 2005].

[177] L. H'walisz, P. Yung, P. Hunggi, P.Talkner, and L. Schimansky-Geier, Zs, f. Phys. **77**, 471 (1989).

[178] K. Mallick and P. Marcq, Eur. Phys. J. B **38**, 99 (2004); ibid. **36**, 119 (2003).

[179] L. Arnold, *Random Dynamic Systems*, [Springer, Berlin, 1998].

[180] L. Tessieri and F. M. Izrailev, Phys. Rev. E **62**,3090 (2000).

[181] K. Mallick and P. Marcq, Phys. Rev. E **66**, 041113 (2002).

[182] P. Hanggi and P. Riseborough, Am. J. Phys. **51**, 347 (1983).

[183] H. K. Leung, Physica A **254**, 146 (1998).

[184] H. K. Leung, Physica A **221**, 340 (1995).

[185] N. Krylov and N. Bogoliubov, Annals Math. Studies **11** [Princeton, 1943].

[186] G. Nicolis, Tellus **34**, 1 (1982).

[187] S.-Q. Jiang, B. Wu and T.-X. Gu, Journal of Electronic Science and Technology **5**, 344 (2007).

[188] S. Jiang, F. Guo, Y. Zhow,and T. Gu, In: *Proceedings of the International Communications, Circuits and Systems Conference*, 11-13 July 2007 pp. 1044-1047.

[189] N. Leprovost, S. Aumaitre, and K. Mallick, Eur. Phys. J. B **49**, 453 (2006).

[190] I. S. Gradstein and I. M. Ryzhik, *Tables of Integrals, Series and Products*, [Academic, Boston, 1994].

[191] V. I. Klyatskin, Radiophys. Quant. Electron. **20**, 381 (1977).

[192] V. Berdichevsky and M. Gitterman, Phys. Rev. E **60**, 1494 (1999).

[193] F. Sasagawa, Progr. Theor. Phys. **69**, 790 (1983).

[194] K. Ouchi, T. Horita and H. Fujisaka, Phys. Rev. E **74**, 031106 (2006).

[195] Y. Jia, X.-P. Zheng, X.-M. Hu, and J.-R .Li, Phys. Rev. E **63**, 031107 (2001).

[196] S. Z. Ke, D. J. Wu and L. Cao, Eur. Phys. J. B **12**, 119 (1999).

[197] V. Braiman and I. Goldhirsch, Phys. Rev. Lett. **66**, 2545 (1991).

[198] Y. Kim, S. J. Lee, S. J.,Kim, Phys. Lett. A **275**, 254 (2000).

[199] J. P. Baltanas, L. Lopez, I. I. Bleckman, P. S. Canda, J. Kurth, and M. A. F. Sanjuan, Phys. Rev. E, **80**, 046008 (2003).

[200] V. N. Chizhevsky and G. Giacomelli, Phys. Rev. A **71**, 011801 (2005).

[201] S. Jeyakumari, V. Chinnathambi, S. Rajasekar, and M. A. F. Sanjuan, Phys. Rev. E **80**, 046608 (2009); Chaos **19**, 043128 (2009).

[202] J. C. Chedjou, H. B. Fotsin, and P. Woafo, Physics Scripta **55**, 390 (1997).

[203] H. G. Schuster and W. Just, *Deterministic Chaos: An Introduction*, [Wiley, 2005].

[204] E. Ott, *Chaos in Dynamical Systems*, [Cambridge University Press, 2002].

[205] P. Berge, Y. Pomeau, and C. Vidal, *Order within Chaos*, [Wiley, New York 1984].

[206] M. Gitterman, Eur. J. Phys. **23**, 119 (2001).

[207] F. Moss, Physics World, February 1977, p. 15.

Index

additive random force, 29
Additive white noise, 18
anomalous diffusion, 24
asymmetric potential, 70

Brownian motion, 24

Coherent stochastic resonance, 76
color noise, 9
correlation functions, 104
critical slowing-down, 128

deterministic chaos, 159
dichotomous noise, 10
diffusion equation, 52
double stochastic resonance, 69
double-well system, 59
Duffing oscillator, 126

effective damping, 98

Floquet-type solutions, 64
fluctuation-dissipation theorem, 4
Fokker-Planck equation, 18
Force-free oscillator, 97
frequency locking, 129
Furutzu-Novikov-Shapiro-Loginov
    procedure, 10

Gaussian white noise, 19
gene selection, 43

Hermit polynomials, 64

internal and external noise, 77
Ito - Stratonovich dilemma, 20

Kapitza pendulum, 51

Langevin equation, 18
limit cycles, 128
linear-response theory, 119

Mathieu equation, 79
matrix continued fraction, 63
Maxwell equations, 90
mean first-passage time, 65
multiplicative noise, 31

noise induced speeding-up, 129
noise-enhanced stability, 48

parametric oscillations, 79
piece-wise linear potential, 49
Poisson noise, 12

random damping, 97
random frequency, 87
random mass, 5
rectangular potential, 52
reentrant transitions, 123
resonant activation, 73

shift of stable points, 31

shot noise, 13
signal-modulated noise, 39
signal-to-noise ratio, 39
single-well potential, 3
statistical linearization, 81
stochastic resonance, 34

trichotomous noise, 11

unified color noise approximation, 23

van der Pol oscillator, 129
vibrational resonance, 156

white noise, 9
white Poisson noise, 12
white shot noise, 13